Vanessa Menezes Costa
Cleto Baratta
Nelson Jorge Batista

Biomonitoring of sewage: genotoxicity and mutagenicity

Vanessa Menezes Costa
Cleto Baratta
Nelson Jorge Batista

Biomonitoring of sewage: genotoxicity and mutagenicity

Sewage treated by stabilization ponds in Teresina-Piauí

ScienciaScripts

Imprint

Any brand names and product names mentioned in this book are subject to trademark, brand or patent protection and are trademarks or registered trademarks of their respective holders. The use of brand names, product names, common names, trade names, product descriptions etc. even without a particular marking in this work is in no way to be construed to mean that such names may be regarded as unrestricted in respect of trademark and brand protection legislation and could thus be used by anyone.

Cover image: www.ingimage.com

This book is a translation from the original published under ISBN 978-613-9-67990-4.

Publisher:
Sciencia Scripts
is a trademark of
Dodo Books Indian Ocean Ltd. and OmniScriptum S.R.L publishing group

120 High Road, East Finchley, London, N2 9ED, United Kingdom
Str. Armeneasca 28/1, office 1, Chisinau MD-2012, Republic of Moldova, Europe
Printed at: see last page
ISBN: 978-620-8-19124-5

TABLE OF CONTENTS

"I can do all things through him who strengthens me." (Philippians 4:13)

ACKNOWLEDGMENTS

I would first like to thank God for all the strength and courage he has given me to complete this master's degree, for all the faith and support I have had in difficult times, my eternal gratitude.

To all my family, especially my mother, Maria do Socorro for her unconditional love, and my aunt Virginia Célia for her example and encouragement during this journey. Thank you all for all your support and trust, you are my foundation.

To my fiancé Jamilson, who was fundamental in this achievement, thank you for your encouragement, strength, patience, dedication and affection.

To my advisor, Professor Dr. Cleto Augusto Baratta Monteiro, for all the help, support and teachings he gave me to complete this research.

My special thanks go to Professor Dr. Nelson Jorge, for all his collaboration, help, encouragement and partnership.

To the company Agua e Esgoto do Piaui S.A (AGESPISA), especially to all the technicians and operators of the East and Pirajà Sewage Treatment Plant for their support, welcome, availability and the training offered to carry out the research.

To the Postgraduate Program in Development and Environment, for the opportunity to develop this work, and to the Federal University of Piaui for all the institutional support.

To all the professors of the program (PRODEMA/UFPI) for the knowledge they passed on, which contributed to making this research great.

To my friends and colleagues, who helped, encouraged and supported me during this journey.

To CAPES for the financial support that helped a lot in carrying out this work.

Thank you all very much!

SUMMARY

Domestic sewage and industrial effluents are the main sources of contamination of water resources, as these wastes are often toxic and their improper disposal can seriously degrade the environment. Polluting agents can cause damage to exposed organisms, including mutagenic damage. Plant tests are often used in environmental monitoring, and the species *Allium cepa* has been widely used to assess the toxic potential of certain environments, especially aquatic ones. The Poti and Parnaiba rivers in the city of Teresina, PI, receive a significant volume of sewage produced in the urban area, as well as being the recipients of effluents processed in sewage treatment plants. The aim of the research was to assess the toxicity and genotoxic and mutagenic potential induced by sewage effluents treated by stabilization pond systems in the city of Teresina - PI using the *Allium cepa* test, as well as to compare the results obtained from the Leste sewage treatment plant, which receives sewage from pit-cleaning vehicles as well as domestic sewage, with those from the Pirajà sewage treatment plant, which receives only domestic sewage. Six samples were collected, two in 2015 and four in 2016, covering the region's dry and rainy seasons at four points: P1-Raw sewage; P2-Upstream; P3-Final effluent and P4-Downstream. The results detected by the phytochemical analyses showed that the water and effluents collected had high concentrations of electrical conductivity, detergents and phosphorus, especially at points P1 and P3, in addition to the high levels of thermotolerant coliforms found in the microbiological analyses at the two sewage treatment plants, the values of which were more critical in the dry season. All the points in the study showed genotoxic and mutagenic potential, as they were able to induce alterations in the genetic material of the test organism used, in addition to a decrease in the mitotic index and inhibition of root growth, mainly evidenced at points P1 and P2 in the dry period, with greater damage identified at the East sewage treatment plant. During the rainy season, the genotoxic and mutagenic potential was lower at both stations compared to the dry season. Therefore, it is possible that the more significant results found at the East sewage treatment plant may have been induced by the release of xenobiotics from other anthropogenic sources, such as pit-cleaning vehicles, which discharge the waste collected at this plant. This waste possibly contains a complex mixture of mutagenic agents, which may be compromising the treatment efficiency of the East sewage treatment plant.

Keywords: Treated effluent; *Allium cepa* test; Genotoxicity; Mutagenicity.

1 INTRODUCTION

Among the environments affected by pollution, one of the hardest hit is the aquatic environment, since water is an essential mineral for the vital functions of organisms and, consequently, for the maintenance of life on the planet. Another worrying factor is the fact that water is a versatile solvent, capable of carrying chemical substances away from sources of contamination, which can compromise other environments (LEME, 2007).

Urban development can compromise the water sustainability of cities and have major impacts due to the load of untreated domestic, industrial and rainwater effluents discharged into reservoirs, along with solid material from waste and erosion, as well as urban growth (LIU; LI, 2010). Domestic sewage and industrial effluents are the main sources responsible for this contamination, as these wastes are often toxic and their presence can seriously degrade the environment (WHITE; RASMUSSEM, 1998).

The impact of discharging effluents from sewage treatment plants into water bodies is a matter of great concern for most countries, and a series of environmental policies and standards are implemented to define criteria for discharge sites and the level of treatment required to ensure that the environmental impacts of disposing of these treated effluents are acceptable (SPERLING, 2005).

All the sewage collected in Teresina is treated in three sewage treatment plants: ETE-Alegria, ETE-Leste and ETE-Pirajà. After proper treatment, this sewage returns to the Poti and Parnaiba rivers (MONTEIRO, 2004). The Poti and Parnaiba rivers receive a significant volume of sewage produced in the urban area of Teresina, as well as being the receiving bodies for the effluent processed in the treatment plants (Stabilization Lagoons). On the right bank of the Parnaiba River, in the Pirajà neighborhood, is the sewage treatment plant - ETE-Pirajà, which serves the Centro/Norte region. The Poti River is the main receiving body for effluents treated by the East Zone Sewage Treatment Plant - ETE-Leste, located on its right bank, as well as effluents from the Alegria Treatment Plant, located on its left bank.

Sewage treatment uses stabilization ponds, which are defined as man-made lentic bodies of water designed to store organic liquid waste, raw and settled sewage, organic and oxidizable industrial waste or oxidized wastewater. Treatment is carried out through natural processes: physical, biological and biochemical, called self-depuration or stabilization (MONTEIRO, 2011).

Despite the efficiency of stabilization pond systems, especially in terms of removing pathogens and organic matter, there is a need for systematic studies to assess the dynamics of the real impact of pond effluents on the Parnaiba and Poti rivers.

Therefore, the *Allium cepa* test developed by Levan (1938) is considered a useful tool for basic research into the genotoxic and cytotoxic potential of chemicals, complex substances such as plant extracts, industrial waste and, above all, contaminated water. This test has been validated internationally as a bioindicator of environmental samples (EVSEEVA *et al.*, 2003).

On the other hand, the *Allium* test is used to assess the quality of bottom, surface and effluent waters, as a simple way of studying macroscopic parameters, both root growth inhibition values and cytological parameters, as well as cell aberrations in metaphases or anaphases and inhibition of dividing cells (FISKESJO, 1988; VESNA *et al.*, 1996; BARBÉRIO *et al.*, 2011).

Studies report that many rivers and reservoirs are contaminated by toxic, genotoxic, mutagenic and carcinogenic substances resulting from the discharge of domestic and industrial effluents (WHITE; RASMUSSEN, 1998), as well as pesticides used in areas adjacent to bodies of water (MONARCA *et al.*, 2000).

The aim of this research is to evaluate the toxicity and genotoxic and mutagenic potential induced by sewage effluents treated by the stabilization ponds of the East and Pirajà Wastewater Treatment Plants using the *Allium cepa* test, as well as to compare the results obtained from the East Wastewater Treatment Plant, which receives not only domestic sewage from the collection network, but also sewage of unknown characteristics transported by pit-cleaning vehicles, with those of the Pirajà Wastewater Treatment Plant, which receives only domestic sewage. In addition to the genotoxic and mutagenic assessment, this study also analyzed the physico-chemical and microbiological parameters of the effluent treated by these sewage treatment plants, obtained from the central laboratory of the company Agua e Esgoto do Piaui S.A (AGESPISA).

In this context, the research question arises: Does the final effluent treated by the stabilization ponds at the treatment plants have genotoxic and mutagenic potential? And does the reception of sewage from pit-cleaning vehicles compromise the efficiency of treatment at the ETE-East and the quality of the water in the Poti River?

This scenario leads to the following study hypotheses:

• It is possible that there is genotoxic and mutagenic potential in the effluent treated at the East Wastewater Treatment Plant, since the composition of the sewage coming from pit-cleaning vehicles is unknown;

• Despite being an unknown sewage, the ETE-East plant has the capacity to treat it;

• The level of genotoxicity and mutagenicity of the ETE-East is higher than that of the ETE-Piraja.

This work is justified by the importance of knowing the degree of genotoxicity and possible mutagenicity present in the sewage effluent treated at the ETE-Leste and Pirajà in the municipality of Teresina, which are discharged into the Poti and Parnaiba rivers, respectively, based on the use of a bioindicator of environmental quality, the *Allium cepa*, a scientifically recognized bioassay, with costs compatible with the simplicity of the method, capable of identifying the toxicity and mutagenic potential of the concentrations of pollutants of organic origin found in the Poti River.

This work is presented in the form of chapters: 1 - Introduction; 2 - General and Specific Objectives; 3 - Theoretical Framework; 4 - Material and Methods; 5 - Results and Discussions; 6 - Conclusion and 7 - References.

2 OBJECTIVES

2.1 . General

To evaluate the toxicity and genotoxic and mutagenic potential induced by sewage effluents treated by stabilization pond systems in the city of Teresina - PI.

2.2 Specific Objectives

•   Analyze the physicochemical and microbiological parameters of sewage effluent samples treated by the ETE-Leste and Pirajà, in accordance with CONAMA resolutions No. 357/2005 and 430/2011;

•   To analyze the genotoxic and mutagenic effects induced by samples of water and sewage treated at four different points in the rainy and dry periods at ETE-East and ETE-Pirajà using the *Allium cepa* test;

•   To compare the results obtained through the *Allium cepa* test of the ETE-East and ETE-Piraja;

3 THEORETICAL FRAMEWORK

3.1 Aquatic pollution

Growing urban and industrial development has had a number of negative impacts on the environment, including pollution of aquatic ecosystems, one of the most affected by anthropogenic action, mainly through the discharge of domestic and industrial effluents without adequate treatment, seriously compromising the environmental quality of this ecosystem.

According to the National Environmental Policy Law No. 6.938/1981, environmental pollution can be defined as the degradation of environmental quality resulting from activities that directly or indirectly harm the health, safety and well-being of the population; create adverse conditions for social and economic activities; adversely affect biota; affect the aesthetic or sanitary conditions of the environment; release materials or energy in disagreement with established environmental standards (BRASIL, 1981). In this way, the disorderly use of natural resources and the poor management of human activities have caused serious impacts on ecosystems, seriously compromising their quality and maintenance (YANG; GUO; SHEN, 2011).

Environmental pollution can occur in both rural and urban areas; by continuous or occasional disposal; by localized sources or by long-distance transport. In this way, the impacts can cause severe alterations to ecosystems and serious damage to living organisms, such as: the accumulation of organic pollutants in the tissues of organisms in places very far from the polluting source, the contamination of surface and underground bodies of water or the contamination of forest soils by pollutants released in industrialized regions (EUGRIS, 2013).

Water pollution is mainly the result of human activities such as the discharge of domestic and industrial effluents without prior treatment. Specific geochemical conditions, such as rainfall and volcanic activity, can also increase the concentration of certain compounds in a given ecosystem and cause local problems (ROSA *et al.*, 2012).

Sources of aquatic pollution can be classified in various ways, taking into account their origin, main components, properties and effects. There are three different categories of pollution: (i) point source discharges where the chemical substances enter the ecosystem in a punctual way, local releases, (ii) chronic discharges in which the discharges occur over time, contaminating a more extensive area and (iii) diffuse releases in which the substances are released over periods (RICHARDSON, 2003).

Among the main chemical pollutants discharged into the aquatic environment are pesticides (TREPÓS *et al.*, 2012), fertilizers (MIRLEAN *et al.*, 2002), industrial chemicals (HENDRYX *et al.*, 2012), polycyclic aromatic hydrocarbons (PAHs) (CELINO *et al.*, *2010*), metals (VENKATRAMREDDY *et al.*, *2009*) and urban effluents (CHEN *et al.*, *2013*), 2010), metals

(VENKATRAMREDDY *et al.,* 2009) and urban effluents (CHEN *et al.,* 2013), with many of these contaminants reaching the aquatic environment through the discharge of effluents that do not receive adequate treatment before being discharged into water bodies (KNIE; LOPES, 2004). According to Liu and Li (2010), in all urban centers, whether they are villages or large cities, the discharge of sewage into our rivers is a constant, with no prospect of solutions.

There are hundreds, perhaps thousands of pollutants that affect the aquatic environment and whose effects are a cause for concern (MATEUCA *et al.,* 2006). The most common aquatic pollutants are pathogens, organic waste, sediments, nutrients and chemical pollutants. These pollutants can spread over the surface and/or through the water column forming "solutions", which can result in undesirable effects on the ecosystem. Organic pollutants such as pesticides, industrial chemicals (polychlorinated biphenyls - PCBs), solvents, detergents and petroleum derivatives are harmful to the aquatic ecosystem (LEGAY *et al.,* 2010).

In addition to the high degree of toxicity of chemical products discharged into aquatic environments, these can also have genotoxic and mutagenic characteristics and can induce alterations in DNA base sequences, increasing the appearance of mutations which, when accumulated, can trigger the appearance of degenerative diseases and neoplastic processes in exposed organisms (RIBEIRO; MARQUES, 2003).

3.2 Urban development, sanitation and effluent discharge

There are approximately 204,450,649 inhabitants in Brazil, with almost 90% of this population living in urban areas, which consequently has an impact on water resources, as investments in basic sanitation in densely populated areas do not exist in the same proportion as the increase in population (IBGE, 2015).

Urbanization not only results in environmental impacts, but also causes the so-called "ecological footprint" in its surroundings, due to the intensive and extensive exploitation of natural resources, large-scale resource extraction and excessive water withdrawal, contributing to the degradation of natural systems with irreversible damage to ecological functions such as the hydrological cycle (MOTA, 2011).

Disordered urban growth generates health risks for the population due to the lack of wastewater treatment and collection services and the disposal of solid waste, which produces a source of internal contamination in the city that helps spread diseases or epidemics, increased risk and frequency of floods, deterioration of the environment, areas degraded by erosion; pollution of rivers and coastal areas, decreasing the recovery capacity of these environments due to high pollutant loads (TUCCI, 2005).

According to Morais (2012), human activities, whether domestic, commercial or industrial, directly or indirectly affect the quality of water bodies, as each type of activity generates specific pollutants which, in certain concentrations in the receiving body, can compromise its use in relation to the expected demands.

In today's globalized world, the growth of economic activities demands greater water supply and sanitation, which ends up putting greater pressure on water resources and natural ecosystems. Faced with this reality, urbanization requires significant investment in water infrastructure for the supply and disposal of wastewater, as a way of preventing polluted and untreated water from posing a risk to public health (TUCCI, 2006).

Today, the inadequate use of water resources is responsible, directly or indirectly, for a series of environmental problems that affect the urban environment and deteriorate the quality of life, especially in peripheral areas, due to the insufficient sewage network, the existence of clandestine connections in the rainwater system and direct discharges into rivers (MAROTTA; SANTOS; ENRICH-PRAST, 2008).

It is clear that as urban development takes place, new anthropogenic activities and demands arise that intensify the production of waste that ends up contaminating watercourses, with the improper dumping of pollutants without proper treatment, affecting all the biodiversity present in these ecosystems.

With regard to the concept of basic sanitation, Law 11.445/07, in its article 3, item I, defines it as comprising the supply of drinking water, sanitary sewage, urban cleaning and solid waste management and urban drainage and stormwater management. In other words, it is the water cycle, from its collection, through its treatment and distribution to the population and ending with final consumption, which includes sanitary sewage - collection and treatment in sewage plants - as well as garbage collection, urban drainage and stormwater management (DUARTE, LAHOZ, 2015).

Nuvolari (2013) refers to basic sanitation as the set of solutions relating to water supply, sewage disposal and solid waste generated, understanding that the appropriate and broader term is "environmental sanitation", understood as a set of actions to preserve the environment and improve the health and quality of life of the population.

On average, the composition of sanitary sewage is 99.9% water and only 0.1% solids, of which around 75% are organic matter in the process of decomposition. In these solids, microorganisms proliferate and pathogenic organisms can occur, depending on the health of the contributing population. These microorganisms come from human feces. Toxic pollutants can also occur, especially phenols and so-called "heavy metals", from mixing with industrial effluents (MONTEIRO, 2011).

Data shows that approximately half of the country's population disposes of their domestic sewage in a sewage or rainwater collection system. More than 20 million Brazilians use septic tanks as a solution for their waste. As a result, Brazil still has a large population without access to these practices and services, with the northeast accounting for almost half of the country's deficit in sewage disposal (BRASIL, 2014).

Brazil serves 48.1% of its population with a sewage system, and only 37.5% of the sewage generated is treated. The Northeast is the second largest region in terms of population, behind only the Southeast, but it turns out to be the second worst region in Brazil in terms of sewage service and treatment. Only 30% of the sewage generated is treated and 21% of the population is served. Teresina, like the state of Piaui, has much lower rates (BRASIL, 2011).

Domestic sewage, whether treated or not, when discharged into a body of water, will cause changes to its physical, chemical and biological characteristics. This change will be greater or lesser depending on the degree of treatment the sewage undergoes, or the level of dilution provided by the receiving body.

Oliveira (2012) points out that the low coverage of sewage treatment in Teresina, the lack of environmental awareness among the population and the lack of investment in preventive actions to prevent pollution of its water bodies, should be given special attention and emphasizes the importance of monitoring the quality of the water in its rivers, as a way of subsidizing preventive actions to control the health of water resources and also of the population.

3.3 Sewage Treatment Plants and Environmental Legislation

A sewage treatment plant (STP) is a set of techniques associated with treatment units, equipment, auxiliary organs (channels, boxes, spillways, pipes) and utility systems (drinking water, fire-fighting, energy distribution, rainwater drainage), the purpose of which is to reduce pollutant loads from sanitary sewage and to condition the residual matter resulting from treatment (MONTEIRO, 2004).

In treatment plants, various operations and unit processes are carried out to separate suspended and dissolved pollutants from the water to be discharged into the receiving body, as well as conditioning the retained waste (NUVOLARI *et al.,* 2003).

Various operations and processes are used in sewage treatment plants to separate suspended and dissolved pollutants from the water to be discharged into the receiving body, one of which is the stabilization pond, as used in Teresina's sewage treatment plants.

These lagoons are bioreactors capable of storing raw sewage, resulting in the stabilization of organic matter through biological processes. Depending on how the organic matter is stabilized, the lagoons can be of the following types: facultative, anaerobic, facultative aerated, decantation and maturation

(MONTEIRO, 2004).

Sperling (2005) describes that the impact of discharging effluents from sewage treatment plants into water bodies is a matter of great concern for most countries. Thus, the establishment of environmental policies and standards is necessary to define criteria for discharge sites and the level of treatment required to ensure that the environmental impacts of the disposal of these treated effluents do not compromise the quality of water resources.

The predominant uses of water resources are established in CONAMA Resolution 357/05, which provides for the classification and environmental guidelines for the classification of surface water bodies and defines thirteen quality classes for fresh, brackish and saline waters in the national territory. With regard to surface water quality, the aforementioned resolution has some descriptive standards for ecotoxicological tests, for example, article 8, item 4, requires that possible interactions between substances and the presence of contaminants not listed therein, which may cause harm to living beings, be investigated using ecotoxicological tests (BRASIL, 2005).

In relation to the conditions and standards for discharging effluents, CONAMA Resolution No. 430 of 2011, which complemented and amended Resolution No. 357 of 2005, states in its article 16 that effluents from any polluting source may only be discharged directly or indirectly into bodies of water after due treatment and provided that they comply with the conditions, standards and requirements laid down in this resolution and in other applicable regulations. The aforementioned resolution states that effluents from any polluting source can only be discharged directly into receiving bodies after due treatment and provided that they comply with the conditions, standards and requirements set out in this Resolution and other applicable norms (BRASIL, 2011).

For a better assessment of environmental samples, which are generally characterized by complex mixtures, it is necessary to carry out more diverse analyses, so that aggression can be estimated more efficiently and quickly and thus enable effective action to be taken to prevent further damage that could promote aggression to the environment (WHO, 2001).

3.4 Genotoxicity and Mutagenicity of water

Most genetic tests look for agents that can affect the genome. Due to the universality of the genetic code, if an agent can cause damage to DNA it has genotoxic potential in any type of cell (animals, plants or micro-organisms) (VILLELA *et al.*, 2003).

Genotoxic agents are those that interact with DNA, altering its structure or function. When these alterations become fixed and acquire the capacity to be transmitted, they are called mutations (UMBUZEIRO; ROUBICEK, 2003).

Mutations are conventionally classified as gene (or point) and chromosomal. Gene mutations refer to changes in one or a few nucleotides of the DNA polymer, through deletions, duplications and/or alterations of base pairs, which therefore end up modifying the functioning of a gene. In chromosomal mutations, there is a reorganization of the DNA structure by translocation, inversion, deletion, duplication, fusion and fission of the chromosomes, altering the chromosomal complement in structure and/or number (JUNDI; FREITAS, 2003).

According to Mitchelmore and Chipman (1998), genotoxicity, the activity of a toxic agent capable of damaging the DNA molecule, is undoubtedly a consequence of environmental pollution. Genotoxic and mutagenic compounds are distributed in aquatic and terrestrial ecosystems (soil and air), being transferred and accumulated, and can cause deleterious effects to exposed organisms (UMBUZEIRO; ROUBICEK, 2003).

Toxicological genetics investigates the mechanisms of action of toxic agents that are capable of inducing highly specific interactions with nucleic acids, which can result in genetic damage, such as point mutation, errors during the DNA replication mechanism, mitotic irregularity, among others (MATSUMOTO; MARIN-MORALES, 2005).

The alterations that occur in the DNA molecule may or may not be correctable by the cell's repair system. When they can be corrected, the promoting agent is called a genotoxic agent, but when these alterations are not corrected by the cell's repair system, the promoting agent is called a mutagen. Thus, alterations in DNA resulting from occasional errors during cell division can result in the appearance of mutations (RIBEIRO; MARQUES, 2003).

With the growth of the human population and industrial production, there has been an increase in polluting discharges, which has led to the deterioration of the environment. Aquatic ecosystems are the most affected, as they end up, in one way or another, serving as temporary or permanent receptors for a huge range of substances capable of damaging the genetic material of exposed organisms and the human population that interacts with this aquatic ecosystem (CLAXTON *et al.,* 1998; ANDRADE *et al.*, 2004; MATSUMOTO *et al.*, 2006; EGITO *et al.,* 2007).

Since aquatic pollution is one of the most alarming, many river water biomonitoring studies have been developed, not only here in Brazil but also in many other parts of the world (MATSUMOTO *et al.,* 2006; PANTALEÂO *et al.,* 2006; SOUZA; FONTANETTI, 2006; EGITO *et al.,* 2007).

3.5 *Allium cepa* test

Environmental biomonitoring means obtaining measurements through some biological organism, either as a whole or through a particular tissue. There are three main situations that lead to biomonitoring: (1) where there are reasons to believe that native species are being threatened; (2) when there are implications for human health regarding the consumption of potentially affected organisms; and (3) when there is an interest in knowing the environmental quality (DA SILVA *et al.*, 2003).

Many plants serve as bioindicators for assessing mutagenic effects and for environmental monitoring, due to their high sensitivity in detecting pollution factors and the toxicity of chemical compounds (LEME; MARIN-MORALES, 2009).

Plant tests are useful for testing complex environmental samples such as sewage (GROVER; KAUR, 1999), river water, reservoirs (RANK; NIELSEN, 1998) and contaminated soils (KOVALCHUCK *et al.*, 1998; COTELLE *et al.*, 1999).

Proof of the efficiency of genotoxic tests carried out on plants for environmental monitoring came from international collaborative studies supported by the United Nations Environmental Program (UNEP), the World Health Organization (WHO) and the *U.S. Environmental Protection Agency* (EPA) (MA *et al.*, 1995). These tests allow the observation of different genetic parameters, such as chromosomal aberrations, which occur from point mutations, both in individual cells and in organs (GRANT, 1994).

The use of higher plants in the diagnosis and monitoring of environmental pollution has been used by several authors. Among these plants, *Allium cepa* (onion) has been used to determine the cytotoxic, genotoxic and mutagenic effects of numerous substances (GRANT, 1994) and complex environmental samples (COTELLE *et al.*, 1999; GROVER; KAUR, 1999; MATSUMOTO; MARINMORALES; 2004; GRISOLIA *et al.*, 2005; EGITO *et al.*, 2007; SOUZA *et al.*, 2008).

Plants of the genus *Allium* have been used to assess the toxicity of effluents and many compounds (ARAMBASIC *et al.*, 1995; RANK; NIELSEN, 1998; TIPIRDAMAZ *et al.*, 2003; TABREZ; AHMAD, 2011). Since the *Allium cepa* test is a eukaryotic system, it can provide a greater degree of proximity when compared to the likely effects on biota exposed to toxic substances, such as those discharged by textile industries, oil spills, etc. (HOSHINA; MARIN-MORALES, 2009, MAZZEO *et al.*, 2011).

Many studies have considered the *Allium cepa* test to be a very effective technique for determining toxicity and pollution levels in the environment (GROVER; KAUR, 1999; CHADRA *et al,* 2005; FATIMA; AHAMAD, 2006; EGITO et al., 2007; LEME; MARIN-MORALES, 2008; CARITÀ; MARIN-MORALES, 2008; LEME; MARIN-MORALES, 2009; BIANCH *et al.*, 2011; MAZZEO *et al.*, 2011; VENTURA-CAMARGO *et al.*, 2011). The test uses the mitotic index (MI) as an indicator of the level of cell proliferation (LEME; MARIN-MORALES, 2009). Reduced or increased values of this parameter may indicate the presence of cytotoxic agents (FERNANDES *et al.*, 2007).

In addition to its sensitivity in detecting cytotoxic, genotoxic and mutagenic effects, the species *A.*

cepa has been indicated as an efficient test organism, thanks to the characteristics it possesses, such as: knowledge of its cell cycle; response to numerous known mutagens; the rapid growth of its roots; the large number of dividing cells; its high tolerance to various cultivation conditions; its availability, due to its easy handling and the fact that it has a small number of chromosomes (2n=16) and a large size, which is a fundamental factor for studies evaluating chromosomal damage and/or cell division cycle disorders, including the risk of aneuploidy (FISKEJO, 1985; QUINZANI-JORDÂO, 1987; GRANT, 1994; EVSEEVA et al., 2003; EGITO *et al.,* 2007; MARIN-MORALES, 2009).

The use of *A. cepa* as a test system was originally introduced by Levan in 1938, when he demonstrated that colchicine could cause disturbances in the mitotic spindle, leading to polyploidization of the meristematic cells of *A. cepa* roots. Later, this same author showed that different solutions of organic salts induced different types of chromosomal aberrations in the meristematic cells of the roots of this plant (LEVAN, 1945).

The use of *Allium cepa* L. (common onion) has been recommended for effluent analysis due to its high sensitivity, low cost, speed, ease of handling and the use of samples without prior treatment, determining the decrease in the mitotic index and the formation of chromosomal aberrations (LEME; MARIN- MORALES, 2009).

Genotoxic substances are capable of inducing damage in parental cells, resulting in chromosomal fragments that are the result of breaks that are not incorporated into the main nucleus of daughter cells after mitosis, defining themselves as small bodies containing DNA and located in the cytoplasm, called Micronuclei (MN) (SCHMID, 1975). In addition to making it possible to analyze chromosomal aberrations, the *Allium cepa* test can also identify the presence of micronuclei.

The Micronucleus (MN) test has been recommended for biomonitoring studies, mainly due to its ability to detect clastogenic agents (chromosome breakage) and aneugenic agents (abnormal chromosome segregation), although it requires cell proliferation to observe the biomarker effect (FENECH, 2000; RIBEIRO *et al.*, 2003).

Micronuclei are small nuclear bodies representing the genetic material that has been lost from the main nucleus as a result of genetic damage. After the chromatids separate in the mitotic process, two nuclei are reconstituted, one at each pole. The nuclear membrane is rebuilt around these two sets of chromosomes. However, if an entire chromosome or an acentric chromosome fragment does not integrate into the new nucleus (because it is not attached to the spindle), it can also form a small individual nucleus (IARMARCOVAI *et al.*, 2007; HOLLAND *et al.*, 2008).

The NM test is an important technique used in environmental monitoring, as it assesses the mutagenic potential of agents present in the environment, and is considered a bioindicator that can evaluate the mutagenicity of the ecosystem (SOUZA; FONTANETTI, 2006). Therefore, the micronucleus test has proved to be a promising *in vivo* and *in vitro* technique for assessing mutagenicity and water quality (ALSABTI; METCALFE, 1995; GRISOLIA; STARLING, 2001).

Bioassays carried out with *A. cepa*, when compared to bioassays carried out with animals, are considered to be more sensitive for environmental assessments and easier to carry out (LEME; MARIN-MORALES, 2008), since the *A. cepa* species has a high germination rate and well-known chromosomal behavior, which allows safe and consistent results to be obtained (ALVIM *et al.,* 2011).

The meristematic cells of *A. cepa* constitute an effective cytogenetic material for analyzing chromosomal aberrations caused by environmental pollution (KRISTEN, 1997), since we can quantify a series of morphological and cytogenetic parameters, including root morphology and growth, determination of the mitotic index, induction of micronuclei and aberrant metaphases, anaphases and telophases (GRANT, 1994; EVSEEVA *et al.,* 2003; EGITO *et al.,* 2007; LEME; MARIN-MORALES, 2009).

The incidence of abnormalities during mitosis in the chromosomes of meristematic cells of *Allium cepa* is an easy method for studying the mechanisms of various genotoxic and mutagenic compounds (KONUK *et al.,* 2007; LEME; MARIN- MORALES 2008; YILDIZ *et al..,* 2009; LIMAN *et al.,* 2010, 2011, 2012; OZKARA *et al.,* 2011); in different environments, such as aquatic (BIANCHI et al. 2011; CARITÀ; MARIN MORALES 2008); as well as in terrestrial habitats and in soil analysis (SOUSA *et al.,* 2009, 2013).

3.6 Physico-chemical parameters

Physico-chemical and microbiological analyses are of great importance for assessing water quality, and these analyses are also important for certifying the results found by the biological tests used in environmental monitoring. Below is a description of the parameters used in the research, together with their importance for assessing water quality.

3.6.1 pH (Hydrogen Potential)

The influence of pH on natural aquatic ecosystems is directly due to its effects on the physiology of different species. The indirect effect is also very important and, under certain pH conditions, can contribute to the precipitation of toxic chemical elements such as heavy metals; other conditions can have an effect on the solubility of nutrients. In this way, pH range restrictions are established for the various classes of natural waters in accordance with federal legislation, the criteria for protecting aquatic life setting the pH between 6 and 9 (CETESB, 2013).

Under standard conditions (25 °C and 1 atm), pH equal to 7 corresponds to neutrality, lower values correspond to the acidic range and values above 7 to the basic (alkaline) range (VILLANUEVA,

2012). It stands out for being a factor that influences a large number of chemical reactions (AGUIRRE-GONZALES *et al.,* 2011), making it an important characteristic to control in a water source, since it influences the biological processes that occur in the aquatic environment, as well as the toxicity of some compounds present in it and the control of physico-chemical processes in the treatment of industrial effluents, since there are many examples of pH-dependent reactions.

In the biological systems formed in sewage treatment, pH is also a condition that has a decisive influence on the treatment process. Normally, the pH condition that corresponds to the formation of a more diverse ecosystem and more stable treatment is neutrality, both in aerobic and anaerobic environments. In anaerobic reactors, the acidification of the environment is indicated by a decrease in the pH of the sludge, indicating a situation of imbalance (CETESB, 2013).

According to Umbuzeiro (2012), pH influences the degree of solubility of various substances, the distribution of free and ionized forms of various chemical compounds, and even defines the toxicity potential of various elements.

3.6.2 Temperature

Temperature plays a crucial role in the aquatic environment, conditioning the influences of a series of physical and chemical variables. In general, as the temperature increases, from 0 to 30°C, viscosity, surface tension, compressibility, specific heat, ionization constant and latent heat of vaporization decrease, while thermal conductivity and vapor pressure increase. Temperature variations are part of the normal climate regime and natural bodies of water show seasonal and diurnal variations, as well as vertical stratification. Surface temperature is influenced by factors such as latitude, altitude, season, time of day, flow rate and depth. The rise in temperature in a body of water is usually caused by industrial discharges (sugarcane industries, for example) and thermoelectric power plants (CETESB, 2013).

The temperature of water and fluids in general informs the magnitude of the kinetic energy of the random movement of molecules and synthesizes the phenomenon of heat transfer to the liquid mass, being directly proportional to the speed of chemical reactions, the solubility of substances and the metabolism of organisms present in the aquatic environment (KULKARNI; CHELLAN, 2010). According to Molnar *et al* (2012), temperature is a factor that influences almost all physical, chemical and biological processes in water

3.6.3 Ammonia

Ammonia (NH_3) is the most reduced form, followed by nitrite ion (NO_2^-) and nitrate ion (NO_3^-). In aerobic environments, the most oxidized form of nitrate ions is likely to be found, while in anaerobic environments, there will be reduced forms such as nitrogen (N_2), ammonia (NH_3) and nitrite ions

(NO$_2^-$) (SHAH; MITCH, 2012). It is the largest nitrogenous residue produced through the metabolism of amino acids and, in water, is reduced to nitrite by bacterial nitrification before being converted to nitrate (VILANUEVA *et al.*, 2012).

Ammonia is a very restrictive toxicant for fish, and many species cannot withstand concentrations above 5 mg/L. In addition, as seen above, ammonia causes dissolved oxygen consumption in natural waters when it is oxidized biologically, the so-called second-stage BOD. For these reasons, the concentration of ammoniacal nitrogen is an important parameter for classifying natural waters and is normally used to compile water quality indices (CETESB, 2013).

Waters with a predominance of organic and ammoniacal nitrogen indicate pollution caused by recent sewage discharges. Nitrates, on the other hand, indicate remote pollution, since they are the end product of nitrogen oxidation. Nitrogen in the form of free ammonia is toxic to fish, and within biochemical processes the conversion of ammonium to nitrite and from this to nitrate consumes dissolved oxygen in the environment, altering the condition of aquatic life (MOULEY *et al.*, 2010; MOLNAR *et al.*, 2012).

3.6.4 Electrical conductivity

Conductivity is the numerical expression of a water's ability to conduct an electric current. It depends on ionic concentrations and temperature and indicates the amount of salts in the water column and therefore represents an indirect measure of the concentration of pollutants. In general, levels above 100 µS/cm indicate impacted environments. Conductivity also provides a good indication of changes in the composition of a water, especially its mineral concentration, but it does not provide any indication of the relative quantities of the various components. The conductivity of water increases as more dissolved solids are added. High values can indicate corrosive characteristics of the water (CETESB, 2013).

According to Yang, Guo and Shen (2011), water has a low ionization potential and therefore small quantities of conductive solutions dissolved in it (inorganic acids, bases and salts) increase its conductivity. Low ionizable solutions such as those formed by organic compounds have low conductivity.

3.6.5 Chlorides

Chloride is the Cl$^-$ anion found in groundwater, which comes from water percolating through soils and rocks. In surface water, sanitary sewage discharges are important sources of chloride, and each person expels around 4 g of chloride per day through urine, which represents around 90 to 95% of human excreta. The rest is expelled through feces and sweat (WHO, 2014).

Chloride in ionic form is one of the main inorganic anions present in water and effluents. It usually comes from the dissolution of minerals or the intrusion of seawater, but can also come from domestic or industrial sewage (chemicals, paints, explosives, matches, paper, electroplating, leather and food processing) in high concentrations, giving the water a salty taste or laxative properties (SHAH; MITCH, 2012).

The concentration of chloride in public water supplies is a standard of acceptance, as it causes a "salty" taste in the water. Concentrations above 250 mg/L cause a detectable taste in the water, but the limit depends on the associated cations. Chloride causes corrosion in hydraulic structures, such as submarine outfalls for the ocean disposal of sanitary sewage, which have therefore been built with high-density polyethylene (HDPE). It also influences the characteristics of natural aquatic ecosystems by causing changes in osmotic pressure in microorganism cells (CETESB, 2013).

3.6.6 Detergents

Detergents are substances that reduce the surface tension of a liquid, and these compounds are also considered surfactants. They are synthetic products produced from petroleum derivatives and began to be produced commercially after the Second World War, due to the scarcity of oils and fats needed to make soap (SILVA *et al.,* 2011).

The main surfactant used in household detergents is anionic, such as alkylbenzene sulfonate (ABS), which was widely accepted on the market because it was more efficient than soap. However, due to the high level of problems in sewage treatment plants, with large foam formations, it had to be replaced by biodegradable surfactants such as linear alkylbenzene sulfonate (LAS) (PENTEADO, 2006).

Because surfactants reduce the surface tension of liquids, it is very easy for bubbles to form and multiply when they are rubbed. When detergents are used that are composed of non-biodegradable surfactants, foam is formed that does not degrade and is immersed in lakes and rivers, causing a great visual impact and damaging the health of aquatic animals, which can lead to their death (ROSA; AFONSO, 2008).

The concentration of detergents in water resources can cause a decrease in dissolved oxygen, due to surface tension; a decrease in light permeability, due to excess foam on the water; and an increase in xenobiotic compounds, chemical compounds that are foreign to an organism or biological system, making it difficult to degrade them (PENTEADO, 2006).

The pollution load resulting from the presence of detergents and soaps is very small compared to other pollutants present in sewage. Despite this, soaps and detergents can make a very unfavorable contribution to water pollution and make wastewater treatment procedures more difficult (ALMEIDA *et al.,* 2003).

3.6.7 Biochemical Oxygen Demand (BOD)

The BOD of a water is the amount of oxygen needed to oxidize organic matter by aerobic microbial decomposition into a stable inorganic form. BOD is normally considered to be the amount of oxygen that aerobic bacteria would need to consume the organic matter present in a liquid (water or sewage). BOD is determined in the laboratory by observing the oxygen consumed in samples of the liquid over 5 days at a temperature of 20 °C (CETESB, 2013).

The most important information provided by this test is the fraction of biodegradable compounds present in the effluent. It is also very important for wastewater treatment work. The BOD test is widely used to assess the pollution potential of domestic and industrial sewage in terms of oxygen consumption. The test is also used for pollution assessment and control, as well as for proposing standards and studies to assess the purification capacity of receiving bodies of water (PESSOA; JORDÂO, 2009).

The greatest increases in BOD in a body of water are caused by discharges of predominantly organic origin. The presence of a high level of organic matter can lead to the complete depletion of oxygen in the water, causing the disappearance of fish and other aquatic life. A high BOD value can indicate an increase in the microflora present and interfere with the balance of aquatic life, as well as producing unpleasant tastes and odors and clogging the sand filters used in water treatment plants (CETESB, 2013).

3.6.8 Chemical Oxygen Demand (COD)

COD corresponds to the amount of oxygen needed to oxidize the organic fraction of a sample that is oxidizable by permanganate or postassium dichromate in acidic solution. The value obtained indicates how much oxygen a given liquid effluent would consume from a receiving body of water after its release, if it were possible to mineralize all the organic matter present, so that high COD values can indicate a high polluting potential (LINS, 2010).

One of the great advantages of COD over BOD is that it allows answers in a shorter time (two hours with dichromate or minutes with specific equipment). In addition, the COD test does not only cover biologically satisfied oxygen demand (as in BOD), but everything that is susceptible to oxygen demand, in particular oxidizable mineral salts (PESSOA; JORDÂO, 2009).

COD values are usually higher than BOD values, and the test is carried out over a shorter period of time. The increase in its concentration in a body of water is mainly due to industrial discharges. It is an indispensable parameter in studies to characterize sanitary sewage and industrial effluents, and is very useful when used in conjunction with BOD to observe the biodegradability of waste (CETESB, 2013).

3.6.9 Total Phosphorus

Phosphorus is an important nutrient for biological processes and its excess can cause water eutrophication. Sources of phosphorus include domestic sewage, due to the presence of superphosphate detergents and fecal matter itself. Rainwater drainage from agricultural and urban areas is also a significant source of phosphorus for water bodies. Industrial effluents include those from the fertilizer, food, dairy, meatpacking and slaughterhouse industries (ANA, 2010).

Phosphorus can appear in water in three different forms. Organic phosphates are the form in which phosphorus makes up organic molecules, such as detergents. Orthophosphates are represented by radicals, which combine with cations to form inorganic salts in water, and polyphosphates, or condensed phosphates, which are polymers of orthophosphates. Like nitrogen, phosphorus is one of the main nutrients for biological processes, in other words, it is one of the so-called macro-nutrients, as it is also required in large quantities by cells. In this capacity, it becomes an essential parameter in programs for characterizing industrial effluents that are to be treated by a biological process (CETESB, 2013).

3.6.10 Dissolved Oxygen (DO)

Dissolved oxygen (DO) is of fundamental importance for the survival of aerobic aquatic beings, such as fish and micro-organisms that use it in their respiration process. It is the most important criterion in determining the health conditions of surface waters. It assesses the effect of oxidizable discharges of organic origin on the water resource, serves as an indicator of the conditions of life in the water and evaluates the process of high purification (HANSEN *et al.,* 2012). It is a parameter measured in the field, and low oxygen concentrations are related to high concentrations of organic matter (BOD), high water temperature, low flow, absence of rapids (USEPA, 1978, 1998, 1999).

The determination of dissolved oxygen is essential for assessing the type of metabolic process that predominates in the aquatic ecosystem. Aerobic conditions favour the aerobic decomposition of organic matter, resulting in a stabilized or mineralized end product and generating odourless and non-toxic products such as CO_2, H_2 and H_2O. An aquatic body with a low oxygen concentration has an unpleasant odour due to the establishment of anaerobic conditions characterized by the fermentation of organic matter (BRASIL, 2005).

3.6.11 Nitrate

Nitrate is found naturally in water and soil in low concentrations. The deposition of organic material in the soil drastically increases the amount of nitrogen. This nitrogen is biochemically transformed and finally becomes nitrate, which is highly mobile in the soil, reaching the underground water source and being deposited there (CAMPOS; ROHLFS, 2011).

Nitrate occurs naturally in groundwater, but its presence in high concentrations is generally the result of anthropogenic activity, particularly the application of organic and inorganic fertilizers and the use of *on-site* sanitation systems. The nitrogenous substances in fertilizers and organic waste are transformed and oxidized by chemical and biological reactions and the result is the presence of nitrate in the soil. As nitrate is extremely soluble in water, it moves easily and contaminates groundwater (BARBOSA, 2005).

High levels of nitrate in drinking water are associated with the occurrence of methemoglobinemia in children. Nitrate causes the oxidation of normal hemoglobin to methemoglobin, which is unable to transport oxygen to the tissues. In adult organisms, these compounds are responsible for high rates of stomach cancer (CETESB, 2013).

3.6.12 Sedimentable Solids (SSed)

In sanitation, solids in water correspond to all the matter that remains as a residue after the sample has been evaporated, dried or calcined at a pre-established temperature for a fixed period of time. In studies to control the pollution of natural waters, especially in studies to characterize sanitary sewage and industrial effluents, determining the concentration levels of the various fractions of solids results in a general picture of the distribution of particles in terms of size (suspended and dissolved solids) and nature (fixed or mineral and volatile or organic) (CETESB, 2013).

For the water resource, solids can cause damage to fish and aquatic life. They can sediment on riverbeds, destroying organisms that provide food or damaging fish spawning beds. Solids can trap bacteria and organic waste at the bottom of rivers, promoting anaerobic decomposition. High levels of mineral salts, particularly sulphate and chloride, are associated with a tendency for corrosion in distribution systems, as well as imparting a taste to the water (CETESB, 2013).

Sedimentable solids (SSed) is the portion of suspended solids that settles under the action of gravity over a period of one hour, from a liter of sample kept at rest in an Imhoff cone. The level of settleable solids in the final effluents discharged by industries is also extremely important as it is a parameter in the legislation. In CONAMA Resolution 357/05, the emission standard is 1 mL/L of settleable solids.

3.7 Microbiological variables

Preserving water quality is a universal need that requires serious attention from the health authorities, and bacteriological tests are essential to assess the quality of the water to be consumed (FATIMA, 2006). Bacteria from the coliform group are the main indicators of fecal contamination and are important as an indicator of the possibility of pathogenic microorganisms (WHO, 2014).

The use of the coliform group and more specifically *Escherichia coli* as an indicator of microbiological quality dates back to its first isolation from feces at the end of the 19th century (WHITE; RASMUSSEM, 1998).

Thermotolerant coliforms are defined as microorganisms of the coliform group capable of fermenting lactose at 44-45°C. They are mainly represented by *Escherichia coli* and also by some bacteria of the *Klebsiella, Enterobacter and Citrobacter* genera. Of these microorganisms, only *E. coli* is exclusively fecal in origin and is rarely found in water or soil that has not been contaminated by fecal matter. The others can occur in water with high levels of organic matter, such as industrial effluents, or in plant material and soil in the process of decomposition. *Escherichia coli* is the main bacterium in the thermotolerant coliform subgroup and is exclusively fecal in origin, considered the most suitable indicator of fecal contamination in fresh waters (CETESB, 2013).

The National Water Agency (2010) points out that thermotolerant coliform bacteria are indicative of domestic sewage pollution and Alves (2011) emphasizes that their presence indicates a risk of other pathogenic microorganisms, responsible for transmitting waterborne diseases.

4 MATERIAL AND METHODS

4.1 Characterization of the study area

The capital of Piauí, Teresina (Figure 1), is located in the CentroNorte Mesoregion of Piauí, between the coordinates 5°08' south latitude and 42°8' west longitude, occupying an area of approximately 1,392 km^2 on the right bank of the Parnaiba River, in the middle course of this Hydrographic Basin, where it receives the Poti River, one of its largest tributaries (IBGE, 2011).

It is Brazil's first planned capital and was officially founded on August 16, 1852. It became the capital due to its central location and the navigability of the Poti and Parnaiba rivers. According to the Teresina Municipal Planning and Coordination Secretariat (SEMPLAM), the capital has 17% of its area considered urban and 83% rural (PMT, 2015).

Figure 1 - Location map of the municipality under study, Teresina - PI

Source: Author, 2016.

The metropolitan area of the capital, Regiao Integrada de Desenvolvimento da Grande Teresina (RIDE), Teresina is made up of the municipalities of Teresina, Altos, Beneditinos, Coivaras, Curralinhos, Demerval Lobao, José de Freitas, Lagoa Alegre, Lagoa do Piaui, Miguel Leao, Monsenhor Gil, Nazària, Pau D'arco and Uniao, in the state of Piaui, as well as the municipality of Timon, which belongs to the state of Maranhao and together totals 1,154,716 inhabitants according to the Demographic Census (IBGE 2011).154,716 inhabitants according to data from the Demographic Census (IBGE, 2011).

The average temperature throughout the year in the municipality of Teresina, as in the whole of the northeast region, changes very little. This phenomenon occurs because the state of Piaui is located close to the equator, where the incidence of solar radiation intensifies the heat over the Teresina region throughout the year (PMT, 2015).

The capital of Piauí, Teresina, is located at the confluence of two important federal rivers, the Parnaiba and the Poti. Its disordered urban growth in recent decades, coupled with ineffective management of its water resources, has left them vulnerable to increased pollution and various impacts that can bring socio-economic and epidemic harm to the local population (OLIVEIRA, 2012).

It has a tropical and rainy (megathermic) savannah climate, with dry winters and rainy summers; the atmospheric air has an annual temperature of 26.8°C and can reach extreme temperatures of 38.6°C, which causes a certain degree of discomfort; it has an average annual rainfall of 1,339mm and relative humidity (annual average) of 70% (MACHADO; PEREIRA; ANDRADE, 2010).

The relief of the municipality has one of the lowest altitudes in the state (100150m), forming a plateau area with flat relief and gentle undulations. The predominant vegetation in Teresina is typical Cerrado, with medium-sized, dense vegetation. Also present in the municipality are forests and coconut groves, which serve as raw material for various activities (PMT, 2015).

4.2 Poti and Parnaiba Rivers

The Parnaiba River is approximately 1,485 km long with its main source located in the Chapada das Mangabeiras under the name of the Agua Quente stream at an altitude of 700 m, between the borders of the states of Tocantins, Maranhao and Piaui. It flows into the Atlantic Ocean, presenting a delta-like mouth where it concentrates around seventy islands, with areas ranging from tens to hundreds of square meters (ARAÙJO, 2006).

The Parnaiba River, the main river in Piauí, is perennial throughout its course and receives contributions from several important tributaries and the water table, from its source to its mouth, and the Poti River, one of the main tributaries of the Parnaiba River, which is considered intermittent, with an average annual flow of 121 m^3 /s (PMT, 2015).

The Parnaiba River Basin, one of the most important basins in the Northeast, covers an area of 331,441 km^2 , of which 249,497 km^2 are in the state of Piauí, 65,492 km in the state of Maranhao, 13,690 km in the state of Ceará and 2,762 km2 are in dispute between Piauí and Ceará (MMA, 2006).

After the São Francisco River basin, the Parnaiba Hydrographic Region is hydrologically the second most important in the Northeast. Its hydrographic region is the most extensive of the 25 basins of the Northeast Strand and covers the state of Piaui and part of the states of Maranhao and Cearà. The region, however, has major inter-regional differences in terms of both economic and social development and water availability. The main tributaries of the Parnaiba are the rivers: Balsas, located in Maranhao; Poti and Portinho, whose sources are in Cearà; and Canindé, Piaui, Uruçui-Preto, Gurguéia and Longa, all in Piaui (ANA, 2015).

The Parnaiba River is one of the largest rivers in the Northeast and plays an important socio-economic role. This is mainly due to the potential of its natural resources, which are suitable for the development of numerous activities: fishing and farming, navigability, electricity, urban supply, leisure, among others (BRASIL, 2006).

The Poti River hydrography is part of the Parnaiba River Basin, the second largest in importance in the Brazilian Northeast region, which represents the densest hydrographic network in this region, covering the entire state of Piaui, which corresponds to 75% of the total area of the basin, lands in the state of Maranhao (19%) and the state of Cearà (6%) (CODEVASF, 2006).

The Poti River Basin is located between the coordinates 4°06' and 6°56' south latitude and 40°30' and 42°50' west longitude, but in Piaui the northern boundary of the basin is at 4°20' south latitude and the eastern boundary is at 40°58' west longitude. The drainage area of the Poti River basin covers 24 municipalities, comprising 52,370 km^2 , of which 38,797 km^2 are located in the state of Piaui. The main tributaries of the Poti River are the Berlangas and Sambito rivers on the left bank and the Canudos and Capivara rivers on the right bank (BRASIL, 2004).

The Poti River Basin, a sub-basin of the Parnaiba River, covers an area of approximately 50,000 km^2 , which corresponds to approximately 16% of the total area of the Parnaiba River Basin. The Poti River is intermittent, with an average annual flow of 121 m^3/s (PMT, 2015).

The river Poti, 550 km long, has its sources in the western region of the state of Cearà, in the Serra dos Cariris Novos, in the town of Jatobà, at an altitude of approximately 800 meters. Like its tributaries, the Poti is a rainfall-dependent river, so its waters change as the rainy season progresses, as is usually the case with the rivers of the Semiarid region (BRASIL, 2004).

In Teresina, the waters of the Poti River are not available to the population for public supply. In relation to the built environment, the establishment of commerce along the avenues along its banks has increased the value of the areas around it. It is important to the population because of the supply of fish, minerals for construction, as a leisure area (bathing) and the use of its banks for agriculture (MENDES-CÂMERA, 2011).

The highest population density in the Poti River basin is in the city of Teresina, which makes it important to seek data and information through the assessment and monitoring of the river's water quality, in order to understand the current conditions and variations in the level of pollution and its effects on the environment and the population (SEMAR, 2004).

The low coverage of sewage treatment in Teresina, coupled with insufficient environmental awareness among the population and little investment in preventive actions against the pollution of its water bodies, draws attention to the importance of monitoring the quality of the water in its rivers, as a means of supporting preventive actions to control the health of water resources as well as the population (OLIVEIRA, 2012).

Improving the quality of the water in the receiving bodies is essential due to its use by the population for leisure activities, given that Teresina has high temperatures throughout the year, due to its climatic characteristics, which establishes a natural attraction for water sports and primary contact recreation with the river (MONTEIRO, 2004).

4.3 Sewage Treatment Plant - ETE

Sewage treatment plants (STPs) have the function of reducing sewage pollution and conditioning the residual matter resulting from treatment. Various operations and processes are used in sewage treatment plants to separate suspended and dissolved pollutants from the water to be discharged into the receiving body, one of which is the stabilization pond, as used in Teresina's sewage treatment plants (PMT, 2015).

Collective sewage treatment in Teresina began in 1969 and was partially completed in 1972, with the implementation of the first stage of the 42-kilometer sewage collection network, which was later extended by a further six kilometers. In 1974, the first stabilization pond was installed at the Pirajà Sewage Treatment Plant (ETE Pirajà) to treat the sewage of 1,200 connections. In 1993, work began on another lagoon at the ETE Leste. In 1995, the Alegria sewage treatment plant was set up in the south of Teresina, and even so, by 1997, the domestic sewage collected and treated corresponded to only 4% of the total number of water connections (PMT, 2015).

The East Sewage Treatment Plant is located on Rua Adalberto Correia Lima, in the Ininga neighborhood. It is located on the right bank of the River Poti, at a minimum distance of 200 meters from residential buildings. It has 15 hectares of treatment ponds, covering a total area of 40 hectares.

The ETE-Leste treats sewage collected in the Iningà, Planalto Uruguai, Piçarreira, Santa Izabel, Morada do Sol, Sao Cristovao, Horto, Jóquei, Fâtima, Noivos neighborhoods, among others, while the ETE Pirajà treats sewage collected in the Pirajà neighborhoods, part of Matadouro, part of the Primavera neighborhood, Matinha, Marquês, Centro, Ilhotas, Cabral, Cristo Rei, Aeroporto, Sao Joaquim, among others (ENGESOFT, 2013).

The East Wastewater Treatment Plant (Figure 2) has a grating system, desanding and stabilization ponds, one of which is an aerated facultative pond, two facultative ponds and two maturation ponds.

It also has a chemical laboratory and a bacteriological laboratory, where effluent analysis is carried out (PMT, 2015).

Figure 2 - East Sewage Treatment Plant, Teresina, PI

Source: MÜLLER, 2004.

The Pirajà Sewage Treatment Plant (Figure 3) is the oldest, located on Avenida Maranhao, on the right bank of the Parnaiba River. It was designed to serve the population of the northern zone and a small part of the central region. Currently, the WWTP has a grating and desanding system and its treatment is carried out using an aerated facultative lagoon and a maturation lagoon (PMT, 2015).

Figure 3 - Pirajà Sewage Treatment Plant, Teresina, PI

Source: MÜLLER, 2004.

It began operating in 1972 with primary treatment using just one lagoon. In 1998, expansion work

28

began, with the construction of a coarse solids separation system, changes to the size of the existing lagoon, the installation of ten aerators and the construction of a maturation lagoon (ROCHA *et al.,* 2001).

The sewage treatment plants, ETE-Pirajà, which discharges its final effluent into the Parnaiba River, and ETE-Leste, which discharges into the Poti River, were designed to receive domestic sewage. However, from 2011 onwards, by decision of the Piaui State Government, the ETE-Leste began to receive sewage transported by Pit Cleaning vehicles, which collect sewage from pits installed in regions that do not receive basic sanitation coverage. As a result, this volume of sewage is now discharged into the ETE-East, so that it is not deposited in the Teresina city landfill.

In 2012, the East Wastewater Treatment Plant received an average of 20 7,000l trucks (a total of 140,000l/day) of septic tank effluent. In 2014 this figure has risen to 45 trucks of 7,000l each, which totals around 315,000l/day. The WWTP has the capacity to receive these trucks, but the problem lies in the organic load coming from these effluents and often the different characteristics they may contain, with compositions different from a simple domestic effluent, since there is no 100% control over the origin of the effluents (PMT, 2015).

4.4 Collection of water and treated effluent samples

The water and effluent samples were collected using 1,500 mL (1.5 L) plastic bottles, with specific preservation for each type of analysis. The procedure for collecting the water samples was carried out in accordance with the APHA (2005) guidelines. In the morning, the plastic bottles were first submerged in the water from the collection site, in order to wash them beforehand with the water from the point to be collected; immediately after this procedure, the samples were collected at the points under study.

After collection, the bottles were immediately sealed for transportation and kept refrigerated until the tests were carried out. A total volume of eight liters of water and effluent was collected and stored in suitable containers for physical-chemical, microbiological and toxicological analysis. Preservation and sampling followed ABNT Standard NBR 9898/ 1987, "Preservation and Sampling Techniques for Liquid Effluents and Receiving Bodies".

4.5 Collection points

The water samples were taken at four different points in the Sewage Treatment Plants (STPs) under study: raw sewage, untreated, upstream, downstream and the final effluent itself, where the treated sewage is discharged into the river (Tables 01 and 02).

Table 1 - ETE-Leste sampling points, Teresina, PI

ETE - LESTE

Point	Name	Latitude	Longitude
P1	Raw sewage - EB	5° 3'5.41"S	42°48'6.71"
P2	Rio Montante - RM	5° 2'43.11"S	42°47'54.27"O
P3	Final Effluent - EF	5° 2'42.18"S	42°47'51.05"O
P4	Downstream River - RJ	5° 2'41.75"S	42°47'47.82"O

Source: Author, 2016.

Figure 4 - Identification of the collection points of the Poti River - ETE-Leste, Teresina, PI

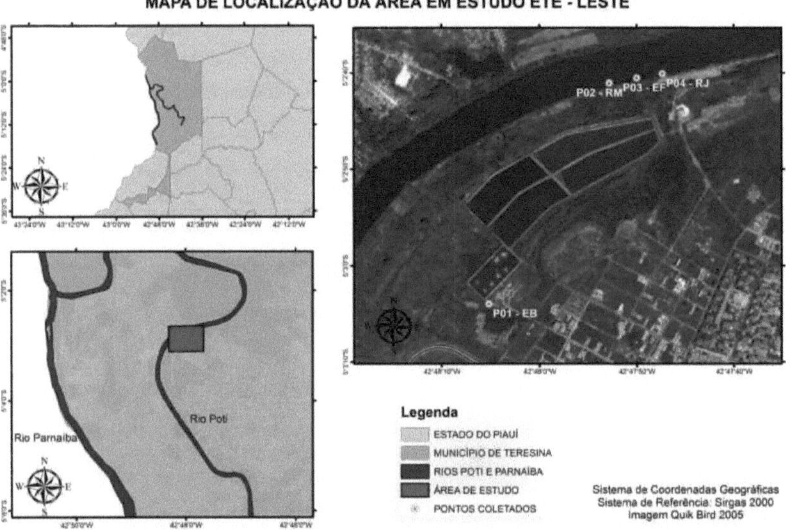

Collections were carried out during two periods of the year to take account of the different climatic conditions in the region: the dry and rainy seasons. Six samples were collected, three in the dry season, November, December 2015 and January 2016, and three in the rainy season, February, March and April 2016.

Table 2 - ETE-Pirajà sampling points, Teresina, PI

	ETE - PIRAJA		
Point	Name	Latitude	Longitude
P1	Raw sewage - EB	5° 4'42.30"S	42°49'45.31"O

P2	Rio Montante - RM	5° 4'31.26"S	42°49'54.50"O
P3	Final Effluent - EF	5° 4'28.31"S	42°49'55.71"O
P4	Downstream River - RJ	5° 4'25.55"S	42°49'57.86"O

Source: Author, 2016.

Figure 5 - Identification of the collection points Parnaiba River - ETE-Pirajâ, Teresina, PI

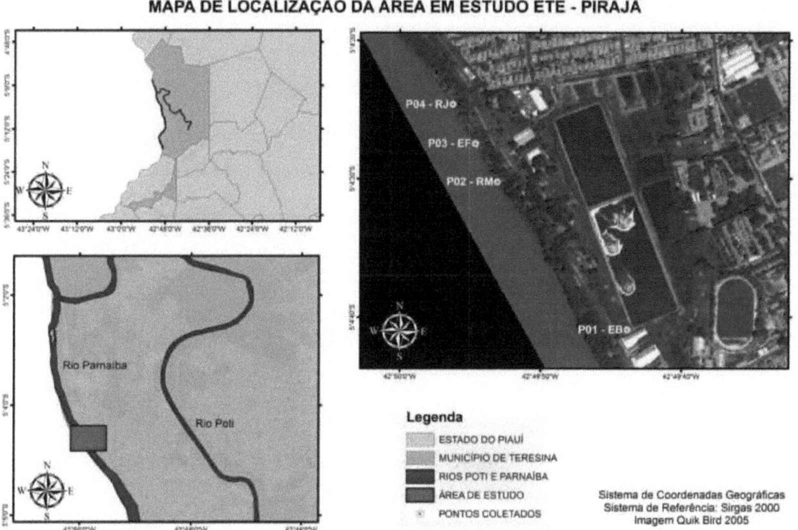

4.6 Analysis of genotoxicity and mutagenicity with *Allium cepa* in water samples and treated effluent

The *Allium cepa* test was carried out at the Fluid Mechanics Laboratory, located at the Technology Center of the Federal University of Piaui-UFPI. It was carried out according to the proposal by Fiskesjo (1993) with some adaptations. The onions used were small, of the same origin, ungerminated and free of pesticides. The toxicity tests used onions of a uniform size, approximately 5.0 cm in diameter, with white, healthy outer cataphylls. The use of small, ungerminated onions of the same origin is recommended. The bulbs were purchased commercially without pesticides and kept in a place free of moisture and protected from light.

Before the experiment, the dried outer cataphylls were removed with a scalpel, making sure that the root area was not damaged. The bulbs were then exposed to tap water for twenty minutes to reduce the effects of possible sprouting inhibitors, and placed to germinate in suitable containers, with the lower part immersed in the test solution or substrate.

Each experiment consisted of 10 (ten) bulbs, as well as a negative control (NC) (water) and a positive control (PC) using a 0.0002 g/L copper sulphate solution (NUNES *et al.,* 2011; BATISTA *et al.,* 2016). The solutions were distributed in previously sterilized glass containers with a capacity of 30 mL. After 72 hours of exposure, the roots were measured using a ruler and those that were too short or too long were discarded.

After measurement, the roots were cut off, 1 cm from the apex, in a total of four roots per bulb, placed in a fixing solution (methanol/acetic acid - 3:1) for 24 hours, then transferred to 70% ethanol and stored in a refrigerator until the slides were prepared histologically.

The test was carried out at a cool temperature of 20° C, on a bench with no vibrations and no direct lighting. To prepare the slides, three or four root tips were removed from the 70% ethanol and washed with distilled water (3 baths of 5 min each) to remove the fixative from the material, then hydrolyzed with 1N HCL for 11 min, followed by another bath in distilled water at room temperature. Then, using fine-tipped tweezers, the roots were dried on filter paper and transferred to dark vials containing Schiff's reagent for approximately 2 hours.

After 2 hours, the roots were washed in running water until the dye was completely removed (the cleaner the wash water, the greater the Schiff's reactivity and the better the coloring of the genetic material). Then the root tips were transferred to a slide, a drop of Fast Green was added to the sectioned material and the slide was placed on top of the slide and squashed with the thumb, with reasonable pressure on the places where the sectioned material is arranged.

The slides were then placed in the refrigerator for around 50 minutes, then glued with Balsam of Canada synthetic glue and the edges of the laminate sealed with enamel so that the material could be reused. The prepared material was then taken to the microscope for observation, and then photographed for better and more efficient reading.

The mutagenic analysis was determined by the mitotic index (MI), the frequency of chromosomal aberrations (CA) in the mitotic cycle and the presence of micronuclei. The mitotic index (MI), which corresponds to the ratio of the number of dividing cells to the total number of cells observed, in percentage, analyzed the presence of metaphase, anaphase and telophase. For the CA analysis, various types of aberrations will be considered (chromosome fragments, chromosome losses, bridges, delays, among others) in the different phases of cell division (metaphase, anaphase and telophase). Toxicity was assessed by measuring the length of the middle roots.

4.7 Evaluation of physico-chemical and microbiological parameters

Analyses of physico-chemical and microbiological parameters were carried out by the physico-chemical and microbiological laboratories of the East Sewage Treatment Plant of the company Aguas e Esgotos do Piaui (AGESPISA). The following parameters were analyzed: temperature, pH, ammonia, electrical conductivity, chlorides, detergents, BOD (biochemical oxygen demand), COD (chemical oxygen demand), total phosphorus, nitrate, DO (dissolved oxygen), and settleable solids. The MPN of total and fecal coliforms was determined using the Colilert technique, which allows the most probable number of microorganisms in the sample to be determined (IDEXX, 2002).

4.8 Evaluation of physico-chemical and microbiological parameters

Statistical analysis was carried out using one-way analysis of variance (ANOVA). The normality of the variables was assessed using the Kolmogorov-Smirnov test. When ANOVA showed significant differences, a *post hoc* analysis was carried out using Tukey's test; when abnormal distribution was observed, comparisons were made using the Kruskal-Wallis test with Dunn's test as *post hoc*. The significance level was 0.05. All analyses were carried out using Prism software, version 5.0.

5 RESULTS AND DISCUSSION

5.1 Analysis of physico-chemical and microbiological parameters

The results obtained for the physicochemical parameters of the samples from the sewage treatment plants and rivers under study during the dry period (November, December 2015 and January 2016) and the rainy period (February, March and April 2016) are shown in Tables 3, 4, 5 and 6. According to CONAMA 357/2005, until the respective classifications are approved, fresh waters will be considered class 2, so the physico-chemical and microbiological analyses of the two rivers under study are based on this classification.

5.2 Sewage Treatment Plants - ETE-Leste and Pirajà

The pH of the samples from both the ETE-Leste and Pirajà analyzed in this study, during the dry and rainy periods, remained within the normal range set out in CONAMA Resolution 430/2011 for freshwater rivers and treated effluent, with values in the range of 6.9 to 7.8.

The temperature variations found for the water and effluent samples collected during the dry and rainy periods remained within the normal range established by CONAMA 430/2011 legislation, varying between 29 and 35°C during the study period. According to Molnar *et al.* (2012), temperature is a factor that influences almost all physical, chemical and biological processes in water.

When comparing the two sewage treatment plants, ETE-Leste and ETE-Pirajà, some of the physico-chemical parameters obtained from the evaluation of the water samples in both collections showed altered results in relation to the current parameters determined for water quality according to CONAMA for class 2 rivers. These changes were evident in the values for ammonia, nitrate, electrical conductivity, detergent, total phosphorus, biochemical oxygen demand (BOD), dissolved oxygen (DO) and settleable solids, especially at points P1 and P3, which refer to the raw sewage and the final (treated) effluent from the treatment plants (Tables 3, 4, 5 and 6).

With regard to the quantification of nitrogenous compounds (ammonia and nitrate), some points showed values above those permitted by the resolution. According to CONAMA 357/2005, the acceptable value for ammonia is up to 20 mg/L and for nitrate ≤ 10 mg/L. Ammonia values were high in P1 and P3 throughout the dry and rainy periods in the two seasons under study. Lower values were observed in the rainy season, and values decreased in P3 after treatment at the station (Tables 3, 4, 5 and 6).

The presence of organic ammonia (NH_3) and inorganic ammonium (NH_4^+) in aquatic environments characterizes recent pollution by domestic sewage, while the presence of nitrate (NO_3^-) characterizes remote pollution, as the nitrogen is in its last stage. The free form of ammonia (NH_3) is toxic but very volatile. Its conversion to nitrite and then nitrate consumes dissolved oxygen, altering the biochemical conditions of the aquatic system. The presence of substances such as ammonia can be a limiting factor for fish, when compared to the recognized toxic effects of this compound on these organisms (CETESB, 2010). In both stations, ammonia values were altered in P1 (raw sewage) and P3 (final effluent). However, in the dry period, the Pirajà WWTP had much lower values than the East WWTP, at points P2 (upstream river) and P4 (downstream river), suggesting that in the Parnaiba River, ammonia levels were possibly lower before and after the discharge of the treated effluent. In the rainy season, however, ammonia levels were lower in the East Wastewater Treatment Plant.

The only altered values were observed in P1 in both seasons, with lower values found during the rainy season. Nitrate is considered a substance with low toxicity, being the end product of nitrification and soluble in water, but it can cause lethal effects in different organisms, or act synergistically with other nitrogenous forms (OSTRENSKY, 1997). Based on the analyses, the nitrate values of the ETE-Leste were higher than those of the Pirajà during the dry period, although after treatment the effluent showed lower values.

According to Tables 3 and 4, during the dry period, electrical conductivity was high at all points in the samples from the ETE-East and at the ETE-Pirajà, those above the maximum permitted value (100 µS/cm - CETESB, 2009) were samples from P1 and P3.

Table 3 - Physico-chemical parameters of water samples and treated effluent from the Poti River (ETE-Leste) in the dry period 2015/2016, Teresina-PI

Physico-chemical parameters of water - RIO POTI

Collection Period	Points	pH	T(⁰C)	Ammonia (mg/L)	EC (µs/cm)	Chlorides (mg/L)	Detergents (mg/L)	BOD5 (mg/L)	COD (mg/L)	Total Phosphorus (mg/L)	OD (mg/L)	Nitrate (mg/L)	Sdm. solids (mL/L/h)
November 2015 (Dry)	Point 1	7,3	31,0	46,5	467,5	90,0	5,0	300,0	485,0	5,7	-	10,7	5,0
	Point 2	6,8	30,0	13,5	350,3	60,0	2,0	20,0	98,0	2,0	0,0	2,7	0,4
	Point 3	7,7	31,0	38,8	477,8	89,0	3,0	97,0	182,0	4,8	6,0	7,6	0,5
	Point 4	6,9	30,0	15,3	334,2	97,0	1,0	39,0	129,0	2,4	0,0	4,2	2,0
December 2015 (Dry)	Point 1	7,3	35,0	47,3	495,4	82,0	8,0	408,0	549,0	7,2	-	13,03	8,0
	Point 2	6,9	29,0	11,6	391,3	87,0	1,2	36,0	165,0	3,4	0,0	5,8	11,0
	Point 3	7,8	31,0	36,0	521,0	120,0	6,0	137,0	233,0	5,3	5,0	8,0	V.A
	Point 4	7,0	30,0	11,8	381,9	89,0	1,4	36,0	195,0	2,9	0,0	4,3	4,0
January 2016 (Dry)	Point 1	7,4	34,0	43,2	472,2	92,0	-	405,0	541,0	5,8	-	-	4,5
	Point 2	6,6	30,0	3,0	261,9	67,0	-	15,0	60,0	0,8	0,0	-	0,8
	Point 3	7,7	31,0	34,6	497,3	153,0	-	115,0	148,0	5,3	6,0	-	-
	Point 4	6,8	30,0	15,5	356,8	78,0	-	45,0	111,0	2,4	1,0	-	3,0
Limits CONAMA 357/430		5 a 9	Until 40°C	≤ 20 mg/L	≤ 100 µs/cm	≤ 250 mg/L	≤ 0.5 mg/L	≤ 120 mg/L	≤ 700 mg/L	0.050 mg/L	≥ 5 mg/L	≤ 10 mg/L	≤ 1.0 mL/L/h

Source: Aguas e Esgotos do Piaui S/A - AGESPISA.

Table 4 - Physico-chemical parameters of water samples and treated effluent from the Parnaiba River (ETE-Pirajâ) in the dry period 2015/2016, Teresina-PI

Physico-chemical parameters of water - RIO PARNAÎBA

Collection Period	Points	pH	T(⁰C)	Ammonia (mg/L)	EC (µs/cm)	Chlorides (mg/L)	Detergents (mg/L)	BOD5 (mg/L)	COD (mg/L)	Total Phosphorus (mg/L)	OD (mg/L)	Nitrate (mg/L)	Sdm. solids (mL/L/h)
November 2015 (Dry)	Point 1	7,4	34,0	44,3	296,9	71,0	3,0	372.0	388,0	6,6	-	10,4	4,0
	Point 2	6,8	30,0	0,6	15,4	1,0	1,0	1,0	31,0	0,4	6,0	0,3	V.A
	Point 3	7,5	30,0	38,8	373,7	67,0	1,0	127,0	240,0	4,8	2,0	9,0	1,1
	Point 4	7,1	30,0	0,6	27,5	2,0	1,0	14,0	39,0	0,1	5,0	0,6	V.A
December 2015 (Dry)	Point 1	7,1	36,0	31,0	367,3	82,0	4,0	288,0	526,0	5,0	-	-	4,0
	Point 2	6,9	32,0	1,2	17,8	7,0	1,0	9,0	90,0	0,2	6,0	-	V.A
	Point 3	7,4	32,0	27,7	389,5	83,0	4,0	127,0	263,0	4,4	3,0	-	0,5
	Point 4	6,9	32,0	2,3	48,5	14,0	1,0	12,0	75,0	0,6	7,0	-	V.A
January 2016 (Dry)	Point 1	7,0	31,0	16,4	185,1	34,0	4,5	210,0	238,0	3,0	-	-	3,0
	Point 2	6,8	29,0	1,0	25,5	1,0	3,0	7,0	15,0	0,5	6,0	-	0,5
	Point 3	7,4	29,0	27,4	332,9	71,0	1,0	171,0	185,0	3,9	3,0	-	V.A
	Point 4	7,2	29,0	3,5	51,5	8,0	3,5	16,0		0,6	5,0	-	0,5
Limits CONAMA 357/430		Up to 5 a to 9	≤ 40° C	≤ 20 mg/L	100≤ µS/cm	250 mg/L	≤ 0.5 mg/L	≤ 120 mg/L	≤ 700 mg/L	0.050 mg/L	≥ 5 mg/L	≤ 10 mg/L	1.0 mL/L/h

Source: Aguas e Esgotos do Piaui S/A - AGESPISA.

It was possible to observe that, in both seasons, the conductivity values decreased in the rainy season, remaining above the permitted level only at points P1 and P3, but it was observed that the values at points P2 and P4 increased in relation to the dry season.

According to Marinelli *et al.* (2000), the increase in electrical conductivity along a river is due to the presence of dissolved materials and can contribute to the impairment of the river's water quality. The high level of electrical conductivity present in the samples from the Poti River is an indication that this spring is impaired by the discharge of effluents, damaging the entire ecosystem present in it. Another factor observed was that the values at P4, on the Parnaiba River, were also higher than at P2 in January, suggesting that after the effluent was discharged, conductivity levels increased.

The electrical conductivity analyses of the samples collected from the river and the sewage treatment plants were relevant to the study, since the ability of a water sample to conduct an electric current is important, as it is directly related to the amount of salts in the water column and therefore represents an indirect measure of the concentration of pollutants. Thus, values recorded above the permitted level also indicate corrosive characteristics of the water (CETESB, 2009) and environments impacted by environmental pollutants (MARINELLI *et al.*, 2000).

From the chemical analyses carried out in this study, it was observed that the chloride values were in line with environmental legislation in both treatment plants during the dry and rainy periods. Chlorides can originate from the dissolution of minerals and soil, the intrusion of saline water, industrial waste or leaching from agricultural areas. They are also present in sewage, due to the contribution of human excreta (PESSOA; JORDÂO, 2009).

With regard to detergents, it was observed that in the dry period most of the points studied had values above those permitted by legislation at the two sewage treatment plants, but the indices were higher at the ETE-East. In the rainy season, the values remained high at points P1 and P3 at both sewage treatment plants, and were lower at points P2 and P4 (Tables 3, 4, 5 and 6).

Sanitary sewage contains between 3 and 6 mg/L of detergents. The indiscriminate discharge of detergents into natural waters leads to aesthetic damage caused by the formation of foam. Detergents have also been blamed for accelerating eutrophication. In addition to the fact that most commercial detergents are rich in phosphorus, they are known to have a toxic effect on zooplankton (PIVELE, 2006).

In the Biochemical Oxygen Demand (BOD) analyses, some points had values above the limit set by CONAMA resolution 430/2011, which is 120 mg/L in the final effluent. The highest levels were found at points P1 and P3 of the two stations, with lower values observed at P2 and P4 of the East Wastewater Treatment Plant during the rainy season (Tables 3, 4, 5 and 6).

As for the levels of Chemical Oxygen Demand (COD), the analyses showed that all the values remained within the established limit, with higher values at points P1 and P3 at both stations during the dry and rainy periods (Tables 3, 4, 5 and 6). The COD test has been used to characterize industrial effluents and to monitor effluent treatment plants in general. Normally the COD of sewage varies between 200 and 800 mg/l (LINS, 2010).

With the BOD and COD analyses of the effluent samples and the rivers under study, it was possible to see that at both stations the values changed were at P1 and P3 for BOD, however, the values at points P2 and P4 of the ETE-Pirajà were lower than in the east during the dry period, and this result can be justified by the fact that the flow and dilution capacity of the Parnaiba River is greater than that of the Poti River.

It was also observed that in the dry period after treatment the BOD value decreased considerably at both stations, and the values at P4 (downstream) were higher after the effluent was discharged. High BOD values mean the presence of a lot of organic matter and a lot of bacteria to break it down, which will compete with other organisms that live in this ecosystem for oxygen in the water, leading to a lack of this compound. COD values were higher than BOD values at all points in the two stations during the dry and rainy months.

Several studies have correlated Chemical Oxygen Demand (COD) and Biochemical Oxygen Demand (BOD$_5$) with water quality levels in Brazil, such as a study presenting water quality indicators in the sub-basin of the Sao Francisco creek in the municipality of Rio Branco - AC - showing significant oxygen oscillation in both the dry and rainy seasons (THEBALDI *et al.*, 2011; SANTI *et al.*, 2012; SILVA *et al.*, 2012).

According to CETESB (2010), when COD values are higher than the values recorded for BOD, this indicates a high load of chemically oxidizable matter which is difficult to biodegrade. The increase in BOD in a body of water is due to the discharge of effluent of predominantly organic origin. According to Guerra (2009), the high content of organic matter in an aquatic environment can induce oxygen molecules to leave the system.

With regard to the total phosphorus analyses, in the dry period the ETE-East had all the values above those established by the environmental resolutions, which establish a maximum permitted value of up to 0.050 mg/L (CONAMA, 2005). In the same period, the Pirajà Wastewater Treatment Plant had points P1 and P3 above the limit. During the rainy season, the phosphorus values were lower and the points with high values were P1 and P3 at both stations, with a reduction in P3 after treatment at the station (Tables 3, 4, 5 and 6).

The presence of excess phosphorus in rivers contributes to the proliferation of algae, which consume a high level of oxygen, reducing oxygen levels in the aquatic environment and causing various negative factors for the ecosystem, such as fish deaths. Phosphorus appears in natural waters mainly due to the discharge of sanitary sewage, industrial effluents and water drained from agricultural and urban areas. The excess of phosphorus present in sanitary sewage and industrial effluents leads to the eutrophication of natural waters (CETESB, 2009). Phosphorus levels above the permitted level can be characterized as pollution due to the dissolution of soil compounds, decomposition of organic matter, domestic and industrial sewage, fertilizers, detergents and animal excrement (MOTA, 2010).

According to CONAMA Resolution 357/2005, Dissolved Oxygen (DO) must be higher than 5 mg/L for class 2 rivers. As far as DO analysis is concerned, it was observed that during the dry period P3 was the only one in the ETE-East to show a dissolved oxygen value (Tables 3, 4, 5 and 6). It is worth noting that during this period the Poti River was highly eutrophicated, caused by high levels of nitrogen and phosphorus, as well as organic matter. The high levels of water hyacinth covered the entire surface of the river, hindering the passage of light and depleting the oxygen needed for aquatic organisms to carry out their functions.

Table 5 - Physico-chemical parameters of water samples and treated effluent from the Poti River (ETE-Leste) during the 2016 rainy season, Teresina-PI

Physico-chemical parameters of water - RIO POTI

Collection Period	Points	pH	T(°C)	Ammonia (mg/L)	EC (µs/cm)	Chlorides (mg/L)	Detergents (mg/L)	BOD5 (mg/L)	COD (mg/L)	Total Phosphorus (mg/L)	OD (mg/L)	Nitrate (mg/L)	Sdm. solids (mL/L/h)
February 2016 (Rainy)	Point 1	7,5	32,0	38,2	487,7	242,0	3,0	480,0	580,0	10,1	-	8,6	20,4
	Point 2	7,0	31,0	0,8	62,7	49,0	0,8	14,0	61,0	0,3	5,0	1,6	V.A
	Point 3	7,5	30,0	30,7	330,5	73,0	24,0	75,0	221,0	4,7	4,0	3,3	2,2
	Point 4	6,8	31,0	0,0	64,1	36,0	0,8	11,0	84,0	0,4	5,0	1,9	ABSENT
March 2016 (Rainy)	Point 1	7,4	32,0	-	400,1	87,0	-	450,0	519,0	4,8	-	11,8	5,0
	Point 2	7,1	30,0	-	51,6	7,0	-	15,0	53,0	1,4	5,0	1,5	ABSENT
	Point 3	7,8	31,0	-	350,9	78,0	-	47,0	61,0	3,5	5,0	2,5	0,3
	Point 4	7,2	30,0	-	60,6	8,0	-	14,0	61,0	0,2	5,0	1,0	ABSENT
April 2016 (Rainy)	Point 1	7,4	32,0	46,1	366,8	118,0	1,8	234,0	400,0	4,9	-	6,7	3,0
	Point 2	7,2	30,0	0,2	59,2	18,0	0,2	16,0	69,0	0,7	5,0	1,9	V.A
	Point 3	7,8	31,0	29,3	343,4	106,0	0,8	36,0	92,0	3,3	6,0	1,3	V.A
	Point 4	7,4	30,0	0,6	63,4	21,0	0,3	15,0	31,0	0,1	5,0	1,9	V.A
Limits CONAMA 357/430		5 a 9	Until 40°C	≤ 20 mg/L	≤ 100 µs/cm	≤ 250 mg/L	≤ 0.5 mg/L	≤ 120 mg/L	≤ 700 mg/L	0.050 mg/L	≥ 5 mg/L	≤ 10 mg/L	≤ 1.0 mL/L/h

Source: Aguas e Esgotos do Piaui S/A - AGESPISA.

Table 6 - Physico-chemical parameters of water samples and treated effluent from the Parnaiba River (ETE-Pirajâ) in the rainy season 2016, Teresina-PI

Period of Collection	Points	pH	T(°C)	Ammonia (mg/L)	EC (µs/cm)	Chlorides (mg/L)	Detergents (mg/L)	BOD5 (mg/L)	COD (mg/L)	Total Phosphorus (mg/L)	OD (mg/L)	Nitrate (mg/L)	Sdm. solids (mL/L/h)
February 2016 (Rainy)	Point 1	7,3	31,0	26,6	258,5	47,0	4,0	420,0	-	3,2	-	-	1,2
	Point 2	7,0	29,0	1,2	29,5	6,0	0,7	24,0	-	2,0	6,0	-	0,2
	Point 3	7,5	30,0	27,0	288,9	61,0	6,0	140,0	-	2,7	2,0	-	V.A
	Point 4	6,9	29,0	0,9	43,6	8,0	0,7	22,0	-	0,1	7,0	-	0,5
March 2016 (Rainy)	Point 1	7,1	31,0	35,3	328,7	58,0	3,0	324,0	382,0	4,3	-	3,8	2,0
	Point 2	7,2	29,0	0,1	34,5	10,0	0,8	8,0	31,0	0,3	8,0	1,7	V.A
	Point 3	7,3	30,0	25,6	284,5	58,0	3,5	167,0	237,0	3,2	2,0	4,0	V.A
	Point 4	7,1	29,0	0,6	73,1	12,0	1,1	12,0	38,0	0,5	7,0	4,4	V.A
April 2016 (Rainy)	Point 1	7,0	32,0	49,3	329,8	80,0	-	330,0	531,0	3,7	-	8,4	2,5
	Point 2	7,4	30,0	0,6	30,6	10,0	-	5,0	31,0	0,4	7,0	1,4	V.A
	Point 3	7,4	30,0	34,0	306,9	87,0	-	137,0	141,0	3,8	2,0	7,0	V.A
	Point 4	7,2	30,0	1,2	63,7	18,0	-	10,0	39,0	0,5	6,0	2,1	0,5
CONAMA 357/430 limits		5 a to 9	Up to 40° C	≤ 20 mg/L	≤ 100 µs/cm	≤ 250 mg/L	≤ 0.5 mg/L	≤ 120 mg/L	≤ 700 mg/L	0.050 mg/L	≥ 5 mg/L	≤ 10 mg/L	≤ 1.0 mL/L/h

Source: Aguas e Esgotos do Piaui S/A - AGESPISA.

In view of these factors, it can be considered as one of the justifications for the zero levels of DO in the Poti River, which indicates that during this period the quality of its waters may have been compromised throughout most of its course, due to the impacts resulting from the discharge of effluents.

At ETE-Pirajà, in the dry period, all the river points, P2 and P4, had adequate DO values, except for P3, which had values below those established in the legislation. In Tables 5 and 6, referring to the rainy season, we can see that the DO values for the East Wastewater Treatment Plant are all in line with what is established, except for P2 in February, and at the Piraja Wastewater Treatment Plant only P3 was below what is established by legislation.

The low concentration of DO found in the water samples is worrying, as it could compromise the maintenance and survival of the aquatic biota of this body of water. Dissolved oxygen is an important physico-chemical parameter for analyzing water quality, as it indicates the capacity of a natural body of water to maintain its endemic biota (CETESB, 2009).

The low DO values found in the samples from the Poti River, below the minimum essential for the maintenance of aquatic life, indicate that the water in this body of water is suffering from environmental imbalance. When dissolved oxygen levels are low, the quality of the water is compromised because there is not enough of it to biodegrade the organic matter (BELLANGER *et al.*, 2004) and may indicate eutrophication due to the discharge of domestic and industrial effluents (MATHEUS; TUNDISI, 1988).

In the quantification of settleable solids, some values above the levels accepted by legislation (1.0 mL/L/h) were observed in both treatment plants during the dry and rainy periods, especially in P1 (Tables 3, 4, 5 and 6).

The amount of settleable matter is an indication of the amount of sludge that can be removed by sedimentation in the decanters. It is an important parameter because it is related to the silting up of the receiving body if its removal is not efficient (LINS, 2010).

5.3 Microbiological tests

With regard to the results of the microbiological analysis of the dry period for the East Wastewater Treatment Plant, it was noted that all the Thermotolerant Coliform values were altered, with indices above what is established by CONAMA 430/2011 legislation ($\leq 4 \times 10^3$). After treating the effluent, there was a decrease in the values. With regard to the results from the Pirajà Wastewater Treatment Plant, only P2 (upstream river) during the period of November and December/2015 had results within the limits allowed by environmental legislation with regard to tolerant coliforms (Tables 7 and 8).

Table 7 - Microbiological analysis of water samples and treated effluent from the Poti River (ETE-Leste) in the dry period 2015/2016, Teresina-PI

Collection Period	Points	Total Coliforms (100 ml)	Thermotolerant Coliforms (*Escherichia coli* - 100ml)
RIO POTI - ETE-LESTE			
	Point 01	$>2,4x10^3$	$3,9x10^7$
November/2015	Point 02	$>2,4x10^3$	$3,1x10^4$
(Dry)	Point 03	$>2,4x10^3$	$1,1x10^5$
	Point 04	$>2,4x10^3$	$3,4x10^4$
	Point 01	$>2,4x10^3$	$2,2x10^7$
December/2015	Point 02	$>2,4x10^3$	$2,4x10^5$
(Dry)	Point 03	$>2,4x10^3$	$3,5x10^4$
	Point 04	$>2,4x10^3$	$1,5x10^4$
	Point 01	$>2,4x10^3$	$2,1x10^7$
January/2016	Point 02	$>2,4x10^3$	$5,3x10^3$
(Dry)	Point 03	$>2,4x10^3$	$1,3x10^5$
	Point 04	$>2,4x10^3$	$3,7x10^4$
Limits CONAMA 430/2011		$\leq 10^3$	$\leq 4x10^3$

Table 8 - Microbiological analysis of samples of water and treated effluent from the Parnaiba River (ETE-Pirajà) in the dry period 2015/2016, Teresina-PI

Collection Period	Points	Total Coliforms (100 ml)	Thermotolerant Coliforms (*Escherichia coli* - 100ml)
PARNAiBA RIVER - ETE-PIRAJA			
	Point 01	$>2,4x10^3$	$2,2x10^7$
November/2015	Point 02	$>2,4x10^3$	$1,1x10^3$
(Dry)	Point 03	$>2,4x10^3$	$3,4x10^5$
	Point 04	$>2,4x10^3$	$7,8x10^3$
	Point 01	$>2,4x10^3$	$2,8x10^7$
December/2015	Point 02	$>2,4x10^3$	$3,5x10^3$
(Dry)	Point 03	$>2,4x10^3$	$2,2x10^5$
	Point 04	$>2,4x10^3$	$2,6x10^4$
	Point 01	$>2,4x10^3$	$1,1x10^7$
January/2016	Point 02	$>2,4x10^3$	$8,2x10^3$
(Dry)	Point 03	$>2,4x10^3$	$7,7x10^4$
	Point 04	$>2,4x10^3$	$1,4x10^4$
Limits		$\leq 10^3$	$\leq 4x10^3$
CONAMA 430/2011			

Source: Aguas e Esgotos do Piaui S/A - AGESPISA.

The preservation of water quality is a universal need that requires serious attention from the health authorities, and bacteriological tests are essential to assess the quality of the water to be consumed (FATIMA, 2006).

Bacteria of the coliform group are the main indicators of fecal contamination and are important as an indicator of the possibility of pathogenic microorganisms (WHO, 1993).

Most of the diseases associated with water are transmitted by fecal pathogens and the control of these waters is through monitoring indicators of microorganisms, since monitoring all the living beings in the water would be a difficult task (LIBÂNIO, 2010).

The most commonly used indicator microorganism is a species of the coliform group *Escherichia coli* (*E. coli*), since it is primarily of fecal origin, rarely able to grow in unpolluted natural environments and is found in large quantities in animal feces (LIBÂNIO, 2010).

In relation to the results of the microbiological analysis of the rainy season for the East Wastewater Treatment Plant, it was noted that only P4 (downstream river) in the period of April/2016 had a result within the limits allowed by CONAMA 430/2011 legislation ($\leq 4x10^3$). As far as the Pirajà WWTP is concerned, only P2 remained within the limits of the aforementioned legislation during the three months of the rainy season. It was also observed that the general values of the microbiological tests increased during the rainy season (Tables 9 and 10).

Table 9 - Microbiological analysis of water samples and treated effluent from the Poti River (ETE-East) during the rainy season of 2016, Teresina-PI

Collection Period	Points	Total Coliforms (100 ml)	Thermotolerant Coliforms (*Escherichia coli* - 100ml)
RIO POTI - ETE-LESTE			
February/2016	Point 01	$>2,4x10^3$	$2,6x10^7$
(Rainy)	Point 02	$>2,4x10^3$	$7,3x10^4$
	Point 03	$>2,4x10^3$	$2,4x10^4$
	Point 04	$>2,4x10^3$	$6,5x10^4$
	Point 01	$>2,4x10^3$	$2,6x10^7$
March/2016	Point 02	$>2,4x10^3$	$1,2x10^4$
(Rainy)	Point 03	$>2,4x10^3$	$8,9x10^4$
	Point 04	$>2,4x10^3$	$1,3 x10^4$
	Point 01	$>2,4x10^3$	$2,2x10^7$
April/2016	Point 02	$>2,4x10^3$	$9,3x10^3$
(Rainy)	Point 03	$>2,4x10^3$	$1,3x10^5$
	Point 04	$>2,4x10^3$	$1,0x10^4$
Limits CONAMA 430/2011		$\leq 10^3$	$\leq 4x10^3$

Table 10 - Microbiological analysis of water samples and treated effluent from the Parnaiba River (ETE-Pirajà) in the rainy season of 2016, Teresina-PI

Collection Period	Points	Total Coliforms (100 ml)	Thermotolerant Coliforms (*Escherichia coli* - 100ml)
PARNAiBA RIVER - ETE-PIRAJA			
	Point 01	>2,4x10³	1,6x10⁷
	Point 02	>2,4x10³	1,9 x10³
February/2016	Point 03	>2,4x10³	3,4x10⁵
(Rainy)	Point 04	>2,4x10³	1,9x10⁴
	Point 01	>2,4x10³	2,8x10⁷
March/2016	Point 02	>2,4x10³	3,2x10³
(Rainy)	Point 03	>2,4x10³	2,7x10⁵
	Point 04	>2,4x10³	2,3x10⁴
	Point 01	>2,4x10³	2,4x10⁷
April/2016	Point 02	>2,4x10³	2,3x10³
(Rainy)	Point 03	>2,4x10³	1,6x10⁵
	Point 04	>2,4x10³	1,8x10⁴
Limits CONAMA 430/2011		≤ 10³	≤ 4x10³

Source: Aguas e Esgotos do Piaui S/A - AGESPISA.

In the microbiological analysis, most of the samples showed levels of thermotolerant and total coliforms well above the permitted levels. The presence of *Escherichia coli* bacteria in a large percentage of the samples is worrying because, as well as not complying with current legislation, this pathogen can involve even lethal cases.

Most of the diseases associated with water are transmitted by fecal pathogens and the control of these waters is through monitoring indicators of microorganisms, since monitoring all the living beings in the water would be a difficult task (LIBÂNIO, 2010).

The most commonly used indicator microorganism is a species of the coliform group *Escherichia coli* (*E. coli*), since it is primarily of fecal origin, rarely able to grow in unpolluted natural environments and is released in large quantities in animal feces (LIBÂNIO, 2010). The monitoring of total coliforms is also carried out with the main objective of verifying the efficiency of disinfection

48

in treatment and the integrity of the water distribution system (BERNARDO; PAZ, 2010). Heterotrophic bacteria can also be used to monitor water quality, especially in distribution systems. The large presence of heterotrophic bacteria in water inhibits the growth of coliform bacteria, which can cause falsely negative results (PADUA; FERREIRA, 2006).

The results observed for the physico-chemical and microbiological standards during the research showed a large contribution of pollutant loads, mostly obtained in P1 (raw sewage), which will always be out of line with the standards, and in P3 (final effluent), from the ETE-Leste and Pirajà, with the ETE-Leste showing higher levels during the dry period, This may have been due to the discharge of unknown effluents from pit-cleaning vehicles at the station, which may be influencing the quality of its treatment, since even after the treatment offered to the effluent, it still returns to the river with high levels of physical, chemical and microbiological compounds, which may be interfering with the quality of the water in this source.

Due to the importance of water quality for health, many toxicity and genotoxicity tests have been used in combination with physical and chemical analyses to assess the quality of water and the environment (SMAKA-KINCL *et al.,* 1996; CLAXTON *et al.,* 1998 , UMBUZEIRO *et al.,* 2001, VARGAS *et al.,* 2001, OHE *et al.,* 2003).

The parameters observed by the physico-chemical analyses point to the need for special attention to be paid to the rivers under study, since the results suggest a possible interference with the biota and the quality of life associated with these environments. High levels of ammonia, phosphorus, DO and BOD can result in changes in the oxygen levels needed to maintain favorable conditions for aquatic life. It is important to note that the analyses of the physical and chemical parameters of the water corroborate the biological results observed in this study using the *Allium cepa* test.

5.4 *Allium cepa* test

The *Allium cepa* test *is* used to assess the quality of bottom, surface and effluent waters, as a simple way of studying macroscopic parameters, both for root growth inhibition values and cytological parameters such as cell aberrations in metaphases or anaphases and inhibition of dividing cells (FISKESJO, 1988; VESNA *et al.,* 1996; BARBÉRIO *et al.,* 2011).

All the samples collected during the dry and rainy periods induced toxicity, as seen by the inhibition of root growth and the inhibition of the mitotic index, as well as mutagenicity, as seen by the induction of micronuclei and chromosomal aberrations (Tables 11, 12, 13 and 14). The positive and negative controls showed the expected results and therefore validate the results obtained in tests with the samples. The results are in line with data in the literature showing a relationship between urban pollution and increased DNA damage (BIANCHI; ESPINDOLA; MARIN-MORALES, 2011;

GEREMIAS *et al.,* 2012).

In the analyses of the *Allium cepa* test carried out with the water and effluent from the Poti and Parnaiba rivers, the action of toxicity was observed in the reduction of the mitotic index (MI) and the reduction in the size of the roots in all the tests, in the different samples collected throughout the dry period. Compared to the positive control (PC) for MI, the most significant values were found at points P1 and P2 for the two sewage treatment plants during the dry period (Tables 11 and 12). In relation to points P3 and P4, the toxicity levels were much lower for both stations when compared to the negative control (NC).

Changes in MI can indicate the presence of cytotoxic agents and are a reliable parameter for evaluating the cytotoxicity of environmental samples and are therefore applicable in environmental monitoring (FERNANDES *et al.,* 2007).

Hoshina (2002), IMs lower than the negative control (NC) may indicate alterations resulting from the action of a given agent on the growth and development of exposed organisms, while IMs higher than those observed in the NC are due to an increase in cell division, which can lead to disordered cell proliferation and even the formation of tumor tissues. Therefore, both the reduction and the increase in MI are important indicators to be considered in environmental contamination assessments.

The analyses of the sizes of the roots of the exposed organisms were consistent with the reduction in MI during the dry season, which shows that the reduction in MI induced a reduction in root growth, especially at points P1 and P2, in the month of November/2015, considered the most critical period (dry season) for all the points, with the most altered values at the ETE-East points (Tables 11 and 12).

In the rainy season, there was an increase in the mitotic index and root length for all sample points. With regard to mutagenicity, chromosome aberrations decreased, along with a reduction in the frequency of micronuclei and failed cytokinesis (binucleated cells) at points P2, P3 and P4 at the two sewage treatment plants (Tables 13 and 14). Point P1 continued to be the most critical in terms of both toxicity and mutagenicity, which was to be expected as it is raw sewage without any treatment, with a high organic and inorganic load.

Chromosomal aberrations (CAs) are alterations characterized by a change in the structure or normal number of chromosomes in a species, and can occur spontaneously or as a result of exposure to chemical or physical agents (RUSSEL, 2002). In this study, mutagenesis was also observed through a significant increase in the frequency of chromosome aberrations (CA) and micronuclei (MN) ($P<0.05$, ANOVA) for all points during the dry period. Differences were observed between the averages of the mutagenic potential of the samples collected during the dry and rainy periods and at the four sampling sites.

Barbério (2009) detected a significant genotoxic effect in the Paraiba do Sul River, with a significant increase in the presence of micronuclei, chromosome bridges, c-metaphase, multipolar anaphase, other chromosome aberrations and the total number of abnormal cells.

The occurrence of micronuclei represents an integrated response of chromosome instability, phenotypes and cellular alterations caused by genetic defects and/or exogenous exposure to genotoxic agents, reflecting numerous chromosomal alterations important for carcinogenesis (FENECH, 2000; FENECH, 2006; FERNANDES *et al.*, 2007).

The most frequent aberrations during the dry season in the Poti River (ETE-Leste) and Parnaiba River (ETE-Pirajà) were anaphase bridges, C-metaphases, binucleated cells and cells with micronuclei, chromosome breaks and delays (Figure 06). During the rainy season, the most frequent aberrations were anaphase bridges and delays and C-metaphases.

Figure 06 - Roots *of A. cepa* grown in samples of water and effluent from the Poti and Parnaiba rivers during the dry season. (A) Metaphase with chromosome breakage; (B and C) C-metaphase; (D) Anaphase with chromosome breakage; (E) Anaphase with delay; (F) Anaphase with bridge; (G) Telophase with delay; (H) Binucleated cell; (I) Cell with micronucleus.

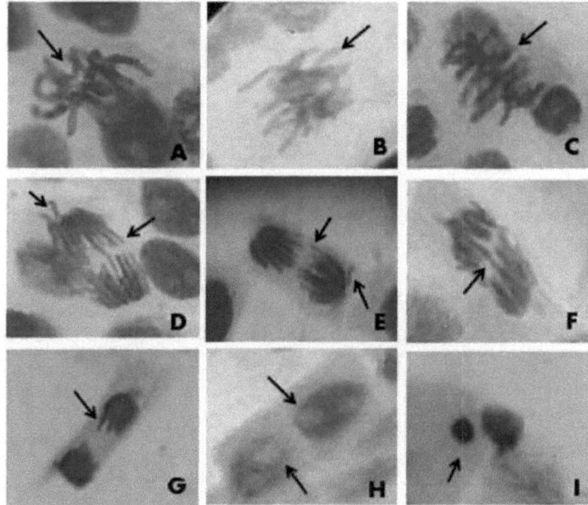

Source: Author, 2016.

Lagging chromosomes are the result of a delay due to a failure to move to one of the poles of the cell. Loose chromosomes can be the result of whole chromosomes of aneugenic origin remaining as laggards. Fragments can come from the action of chemical agents that induce chromosome breaks (clastogens) and their interference in the chromosomes is associated with breaks in the DNA molecule

(TURKOGLU, 2008).

According to Fiskesjo (1993), chromosome breaks can cause fragments and anaphase bridges, the latter being caused by translocations or simply by cohesive ends. An anaphase bridge can result, according to Hall (1994), from breaks that occur in the G_2 period of the cell cycle, after chromosome replication.

Micronuclei (MN) originate from acentric or late fragments, which are excluded from the nucleus itself during mitosis (HOLLAND *et al.,* 2011; GUPTA; AHMAD, 2012). The micronucleus test detects mutagenesis in eukaryotic organisms, such as clastogenesis, aneugenesis and damage to the mitotic spindle (SILVA *et al.,* 2011).

Anaphase bridges consist of chromatid bridges between chromosome portions that separate during anaphase. These bridges, when broken, can result in the loss of genetic material, such as micronuclei (FISKESJO, 1985; FISKESJO, 1993; NIELSEN, 1994; MATSUMOTO; MARIN-MORALES, 2004).

In this study, all the samples, except for point 04 (downstream river), collected during the dry and rainy periods, induced toxicity, verified by the inhibition of root growth and inhibition of the mitotic index, but also induced mutagenicity, evidenced by the induction of micronuclei and chromosomal aberrations. Our results are consistent with those found in the literature showing the relationship between urban pollution and increased DNA damage (BIANCHI *et al.,* 2011; GEREMIAS *et al.,* 2012; TABET *et al.,* 2015).

The results presented in tables 11 and 12 show that the points, P1 (raw sewage) and P2 (upstream river), showed significant mutagenic activity (AC and MN) throughout the dry period, compared to the CP in both the ETE-East and ETE-Pirajâ.

The increase in mutagenicity in P2 suggests that the Poti and Parnaiba rivers may possibly be receiving effluents before receiving treated effluent from sewage treatment plants, especially the Poti river, where the values found were more significant.

According to Fenech (2000), contaminants are capable of inducing chromosomal aberrations by various mechanisms, with clastogenic and aneugenic action being the most commonly observed. Clastogenic action is characterized by chromosome breakage during cell division, and aneugenic action occurs when an agent promotes the inactivation of a cytoplasmic structure, for example, the mitotic spindle, leading to the loss of one or more chromosomes.

Table 11 - Mitotic index, chromosomal aberrations, micronuclei and root length in *A. cepa* exposed to water and treated effluent from the Poti River (ETE-Leste) in the dry period 2015/2016, Teresina-PI

Period of Collection (Dry)	Group	Mitotic index (dividing cells/2000)	Chromosomal aberrations				MN/ 2000	Binucleated/ 2000	Root length (cm)
			Anaphasic Bridges	Chromosome fragments	Anaphasic Delays	C-metaphases			
	Control Negative[a]	434,3 ± 1,70	0,30 ± 0,71	0,26 ± 0,54	0,80 ± 0,21	0,12 ± 0,28	0,23 ± 0,47	0,26 ± 0,42	7,62 ± 1,26
November 2015	Point 01	42,1 ± 17,78***	3,60 ± 1,42	1,80 ± 0,42	3,60 ± 1,43**	0,80 ± 0,82	4,30 ± 0,67***	1,25 ± 0,33	2,64 ± 1,42***
	Point 02	52,40 ± 31,46***#	2,80 ± 1,21	1,10 ± 0,5*#	2,40 ± 1,26***	0,60 ± 0,42	3,80 ± 0,24***	0,92± 0,34	4,43 ± 1,72***#
	Point 03	228,80 ± 12,24***#	0,82 ± 0,24***#	0,62 ± 0,32	1,83 ± 1,21***	0,43 ± 0,23	1,64 ± 0,56***£	0,74 ± 0,32***#£	5,36 ± 1,34***#
	Point 04	342,80 ± 14,26***#	0,68 ± 0,36***#	0,52 ± 0,38	0,94 ± 1,21***	0,28 ± 0,26	0,82 ± 0,24***£	1,52 ± 0,37***#£	5,62 ± 1,13***#
December 2015	Point 01	58,1 ± 14,26***	2,13 ± 1,48	1,61 ± 0,14	3,42± 1,26**	0,78 ± 0,23	3,88 ± 0,36***	1,13 ± 0,15	3,36 ± 1,14***
	Point 02	65,30 ± 23,14***#	2,26 ± 1,14	1,08 ± 0,4*#	2,64 ± 1,24***	0,54 ± 0,12	3,74 ± 0,16***	0,72± 0,23	4,67± 1,16***#
	Point 03	352,18± 14,32***#	0,92 ± 0,36***#	0,54 ± 0,23	1,52 ± 1,14***	0,32 ± 0,21	1,51 ± 0,17***£	0,63 ± 0,31***#£	5,84 ± 1,23***#
	Point 04	383,25± 16,23***#	0,43 ± 0,13***#	0,36 ± 0,12	0,88 ± 1,13***	0,21 ± 0,13	0,78 ± 0,13***£	0,58 ± 0,24***#£	6,04 ± 1,28***#
January 2016	Point 01	62,9 ± 15,12***	1,64 ± 0,51	1,23 ± 0,26*	3,88 ± 1,28***	0,64 ± 0,42	3,63 ± 1,23***	1,08 ± 0,62	4,82 ± 0,34***
	Point 02	72,24 ± 7,27***#	1,53 ± 0,25*	1,14± 0,32***#	3,25 ± 1,32***	0,52 ± 0,12	2,73± 0,61***	1,02 ± 0,52***#	5,05 ± 0,25***
	Point 03	364,36 ± 4,14***#	0,74 ± 0,26***#£	0,43 ± 0,31***#	1,86 ± 1,31***	0,30 ± 0,18***£	3,24 ± 1,14***	0,52 ± 0,24***#	5,83 ± 0,48***
	Point 04	388,24 ± 6,13***#	0,42 ± 0,12***#£	0,35 ± 0,13***#	0,76 ± 1,23***	0,23 ± 0,32***£	2,86 ± 1,21***	0,46 ± 0,23***#	6,25± 0,43***
	Positive Control[b]	32,70 ± 13,1***	2,60 ± 0,42*	0,90 ± 0,10**	2,30 ± 0,72*	0,60 ± 0,41	5,40 ± 0,62***	1,70 ± 0,32***	1,88 ± 1,2***

· Negative Control= well water;· Positive Control= copper sulphate (0.0002 g/L); * Significantly different from the negative control at the level of P<0.05; ** P<0.01; *** P<0.001 (ANOVA); #

Significant difference from Point 1 at P<0.05 (ANOVA); £ Significant difference from Point 2 at P<0.05 (ANOVA).

Table 12 - Mitotic index, chromosomal aberrations, micronuclei and root length in *A. cepa* exposed to water and treated effluent from the Parnaiba River (ETE-Pirajà) in the dry period 2015/2016, Teresina-PI

Period of Collection (Dry)	Group	Mitotic index (dividing cells/2000)	Anaphasic Bridges	Chromosome fragments	Anaphasic Delays	C-metaphases	MN/2000	Binucleated/2000	Root length (cm)
	Control Negative[a]	458,4 ± 1,82	0,26 ± 0,13	0,18 ± 0,14	0,62 ± 0,13	0,18 ± 0,24	0,28 ± 0,36	0,21 ± 0,32	7,14 ± 1,38
November 2015	Point 01	56,4 ± 13,24***	3,20 ± 1,34	1,34 ± 0,61	3,23 ± 1,16**	0,76 ± 0,52	4,08 ± 0,34***	1,14 ± 0,42	2,53 ± 1,32***
	Point 02	52,84 ± 22,36***#	2,21 ± 1,14	1,02 ± 0,6*#	2,04 ± 1,53***	0,48 ± 0,87	3,20 ± 0,43***	0,86 ± 0,26	3,56 ± 1,47***#
	Point 03	284,32 ± 15,16***#	0,78 ± 0,82***#	0,53 ± 0,25	1,12 ± 1,35***	0,33 ± 0,12	1,21 ± 0,32***£	0,61 ± 0,25***#£	5,45 ± 1,13***#
	Point 04	361,16 ± 11,28***#	0,61 ± 0,26***#	0,43 ± 0,21	0,82 ± 1,52***	0,20 ± 0,14	0,78 ± 0,25***£	1,31 ± 0,26***#£	5,82 ± 1,28***#
December 2015	Point 01	64,4 ± 12,26***	2,05 ± 1,32	1,22 ± 0,17	3,15 ± 1,52**	0,63 ± 0,13	3,21 ± 0,96***	1,06 ± 0,13	4,84 ± 1,34***
	Point 02	60,14 ± 18,16***#	2,15 ± 1,35	0,98 ± 0,6*#	1,82 ± 1,42***	0,42 ± 0,11	3,34 ± 0,75***	0,61 ± 0,25	5,92 ± 1,63***#
	Point 03	370,15 ± 10,16***#	0,81 ± 0,63***#	0,41 ± 0,32	1,21 ± 1,32***	0,30 ± 0,72	1,21 ± 0,37***£	0,51 ± 0,17***#£	5,98 ± 1,35***#
	Point 04	388,57 ± 13,23***#	0,38 ± 0,47***#	0,28 ± 0,8	0,75 ± 1,73***	0,18 ± 0,32	0,66 ± 0,74***£	0,42 ± 0,86***#£	6,26 ± 1,52***#
January 2016	Point 01	78,5 ± 11,27***	1,27 ± 0,31	1,06 ± 0,13*	3,52 ± 1,73***	0,52 ± 0,83	3,02 ± 1,37***	1,04 ± 0,93	5,08 ± 0,46***
	Point 02	71,15 ± 6,42***#	1,48 ± 0,13*	0,87 ± 0,26***#	3,12 ± 1,22***	0,42 ± 0,63	2,04 ± 0,63***	1,02 ± 0,57***#	5,43 ± 0,54***
	Point 03	375,48 ± 6,64***#	0,51 ± 0,64***#£	0,38 ± 0,92***#	1,41 ± 1,32***	0,26 ± 0,62***£	3,12 ± 1,84***	0,44 ± 0,15***#	6,46 ± 0,21***
	Point 04	372,56 ± 6,82***#	0,36 ± 0,82***#£	0,31 ± 0,7***#	0,63 ± 1,74***	0,17 ± 0,41***£	2,11 ± 1,53***	0,38 ± 0,26***#	6,85 ± 0,14***
	Control Positive[b]	38,25 ± 11,13***	3,40 ± 0,26*	0,84 ± 0,14**	3,25 ± 0,83*	0,76 ± 0,32	6,24 ± 0,43***	2,47 ± 0,41***	1,46 ± 1,6***

[a] Negative Control= dechlorinated water; [b] Positive Control= copper sulphate (0.0002 g/L); * Significantly different from negative control at $P<0.05$; ** $P<0.01$; *** $P<0.001$ (ANOVA); # Significantly different from Point 01 at $P<0.05$ (ANOVA); £ Significantly different from Point 02 at $P<0.05$ (ANOVA).

The increase in mutagenicity in P2 during the dry season suggests that the Poti and Parnaiba rivers may possibly be receiving effluent before receiving treated effluent from sewage treatment plants, especially the Poti river, where the values found were more significant (Tables 11 and 12).

Thomas *et al.* (2009) suggest that the number of binucleated cells may be an important biomarker of cytokinesis failures and may cause a higher proportion of aneuploidy. The presence of such anomalies is indicative of high levels of toxicity that promote irreversible damage to the cell, thus leading to cell death (TÜRKOGLU, 2012).

The frequency of AC and MN in the analyses decreased after the effluent was treated at the stations. Points P3 and P4 had values close to CN, resulting in a low mutagenic potential during the dry and rainy periods (Tables 11, 12, 13 and 14). After treatment, the genotoxic and mutagenic potential suffered a reduction in toxicity levels, possibly reducing the impact on river waters after receiving the effluent.

High concentrations of xenobiotics in urban areas, agricultural and industrial waste have contributed to an increase in genotoxic activity in aquatic environments. It is now known that one of the main sources of pollution in freshwater aquatic ecosystems is urban solid waste (VILLELA *et al,* 2003).

Factors such as seasonality, rainfall and flow can influence the concentration of pollutants in water (OLIVEIRA *et al.,* 2011). Some studies, such as on the Paraiba do Sul River in Sao Paulo, suggest the influence of seasonality on the induction of genotoxic effects (MN) (LEMOS; SOUZA; FONTANETTI, 2006). Seasonal variation is a factor that can significantly influence the frequency of genetic damage and promote physiological changes in exposed organisms (OLIVEIRA *et al.,* 2011).

However, during the rainy season the induction of genotoxic and mutagenic activity was lower, which is believed to be due to the dilution of organic and inorganic loads at this station. There was a reduction in micronuclei and binucleated cells at both stations, except at point 01, considered the most critical. There was also a reduction in mutagenicity at point 02 (upstream), which was very critical during the dry season (Tables 13 and 14).

Table 13 - Mitotic index, chromosomal aberrations, micronuclei and root length in *A. cepa* exposed to water and treated effluent from the Poti River (ETE-Leste) in the rainy season 2016, Teresina-PI

Period of Collection (Dry)	Group	Mitotic index (dividing cells/2000)	Chromosomal aberrations				MN/2000	Binucleated/2000	Root length (cm)
			Anaphasic Bridges	Chromosome fragments	Anaphasic Delays	C-metaphases			
	Control Negative[a]	486,3 ± 2,6	0,24 ± 0,42	0,21 ± 0,23	0,68 ± 0,13	0,9 ± 0,3	0,18 ± 0,4	0,18 ± 0,6	8,35 ± 1,8
November 2015	Point 01	52,3 ± 2,24***	2,63 ± 1,36	1,6 ± 0,3	2,8 ± 1,12**	0,6 ± 0,3	3,24 ± 0,32***	1,13 ± 0,32	4,25 ± 1,14***
	Point 02	64,21 ± 22,12***#	2,4 ± 1,2	0,96 ± 0,14*#	1,82 ± 1,11***	0,54 ± 0,13	2,96 ± 0,34***	0,82 ± 0,21	6,14 ± 1,62***#
	Point 03	284,13 ± 11,42***#	0,74 ± 0,21***#	0,43 ± 0,13	1,23 ± 1,3***	0,36 ± 0,15	1,26 ± 0,14***£	0,66 ± 0,31***#£	7,23 ± 1,12***#
	Point 04	396,22 ± 18,34***#	0,52 ± 0,23***#	0,48 ± 0,25	0,82 ± 1,8***	0,22 ± 0,11	0,73 ± 0,13***£	1,21 ± 0,26***#£	7,94 ± 1,42***#
December 2015	Point 01	65,13 ± 11,16***	1,86 ± 1,43	1,32 ± 0,16	2,85 ± 1,26**	0,66 ± 0,13	2,64 ± 0,26***	1,12 ± 0,6	5,26 ± 1,51***
	Point 02	72,27 ± 12,42***#	1,74 ± 1,25	1,11 ± 0,8	1,52 ± 1,62***	0,46 ± 0,16	2,26 ± 0,8***	0,52 ± 0,12	6,92 ± 1,43***#
	Point 03	336,23 ± 12,48***#	0,63 ± 0,25***#	0,32 ± 0,16	1,13 ± 0,7***	0,21 ± 0,9	1,18 ± 0,9***£	0,56 ± 0,24***#£	7,42 ± 1,33***#
	Point 04	362,12 ± 14,34***#	0,35 ± 0,8***#	0,24 ± 0,7	0,6 ± 0,3***	0,16 ± 0,8	0,62 ± 0,16***£	0,42 ± 0,12***#£	7,62 ± 1,15***#
January 2016	Point 01	68,63 ± 13,25***	1,24 ± 0,12	1,11 ± 0,4*	2,32 ± 1,14***	0,34 ± 0,12	2,14 ± 1,8***	1,2 ± 0,6	6,25 ± 0,28***
	Point 02	78,12 ± 8,32***#	1,12 ± 0,6*	1,5 ± 0,5***#	2,54 ± 1,26**	0,42 ± 0,8	2,14 ± 0,33***	1,3 ± 0,3***#	7,33 ± 0,54***
	Point 03	288,24 ± 5,27***#	0,38 ± 0,14***#£	0,15 ± 0,7***#	1,28 ± 1,4***	0,12 ± 0,6***£	2,16 ± 1,4***	0,42 ± 0,14***#	7,64 ± 0,84***
	Point 04	372,28 ± 7,25***#	0,25 ± 0,5***#£	0,24 ± 0,7***#	0,61 ± 1,12***	0,18 ± 0,7***£	1,18 ± 1,7***	0,25 ± 0,12***#	7,62 ± 0,88***
	Control -. -b	46,25 ± 16,4***	2,42 ± 0,12*	0,64 ± 0,8**	1,88 ± 0,63*	0,52 ± 0,34	4,24 ± 0,32***	1,21 ± 0,24***	1,94 ± 1,8***

· Negative control = dechlorinated water;· Positive control = copper sulphate (0.0002 g/L); * Significantly different from negative control at P<0.05; ** P<0.01; *** P<0.001 (ANOVA); #

Significant difference from Point 1 at P<0.05 (ANOVA); £ Significant difference from Point 2 at P<0.05 (ANOVA).

Table 14 - Mitotic index, chromosomal aberrations, micronuclei and root length in *A. cepa* exposed to water and treated effluent from the Parnaiba River (ETE-Pirajà) in the rainy season 2016, Teresina-PI

Period of Collection (Dry)	Group	Mitotic index (dividing cells/2000)	Chromosomal aberrations				MN/2000	Binucleated/2000	Root length (cm)
			Anaphasic Bridges	Chromosome fragments	Anaphasic Delays	C-metaphases			
November 2015	Control Negative[a]	472,5 ± 1,25	0,19 ± 0,4	0,12 ± 0,6	0,52 ± 0,12	0,8 ± 0,4	0,22 ± 0,8	0,16 ± 0,8	7,63 ± 1,53
	Point 01	68,5 ± 16,13***	2,1 ± 0,9	1,26 ± 0,13	2,02 ± 1,8**	0,26 ± 0,8	3,33 ± 0,15***	1,2± 0,6	5,34 ± 1,55***
	Point 02	72,24 ± 33,13***#	2,06 ± 1,32	0,9 ± 0,3*#	1,68 ± 1,12***	0,32 ± 0,13	2,56 ± 0,36***	0,52± 0,26	6,27 ± 1,53***#
	Point 03	294,26 ± 18,26***#	0,28 ± 0,12***#	0,32 ± 0,11	1,06 ± 1,2***	0,18 ± 0,9	1,6 ± 0,3***£	0,52± 0,18***#£	7,68 ± 1,64***#
	Point 04	322,12± 14,43***#	0,32 ± 0,14***#	0,26 ± 0,13	0,46 ± 1,4***	0,12 ± 0,6	0,42 ± 0,12***£	1,21 ± 0,7***#£	7,93 ± 1,74***#
December 2016	Point 01	76,5 ± 16,32***	1,43 ± 1,13	1,14± 0,7	2,86± 1,16**	0,38 ± 0,4	2,45 ± 0,13***	0,86 ± 0,43	6,24 ± 1,16***
	Point 02	83,14± 19,12***#	1,26 ± 1,3	0,64± 0,8*#	1,36 ± 1,14***	0,24 ± 0,7	3,12± 0,8***	0,48± 0,12	6,88± 1,52***#
	Point 03	298,12± 11,24***#	0,43 ± 0,15***#	0,23± 0,7	1,14 ± 1,7***	0,12 ± 0,6	1,08± 0,4***£	0,23 0,6***#£	7,46 ± 1,63***#
	Point 04	314,24± 11,26***#	0,27 ± 0,14***#	0,16± 0,4	0,62 ± 1,14***	0,11 ± 0,5	0,28 0,12***£	0,26 0,13***#£	7,84 ± 1,72***#
January 2016	Point 01	82,23± 15,22***	1,16 ± 0,8	0,84 ± 0,8*	2,26 ± 1,14***	0,38 ± 0,23	2,56 1,22***	0,96 ± 0,13	6,93 0,85***
	Point 02	96,22 ± 8,32***#	1,16± 0,8*	0,52± 0,14***#	2,35 ± 1,8***	0,18 ± 0,8	0,76± 0,42***	0,26 0,7***#	7,24 ± 0,73***
	Point 03	276,23 ± 5,32***#	0,46 ± 0,14***#£	0,21 ± 0,7***#	1,14± 1,7***	0,12 ± 0,6***£	2,58± 1,26***	0,36± 0,13***#	7,32 ± 0,58***
	Point 04	316,36 ± 7,43***#	0,15 ± 0,6***#£	0,26± 0,9***#	0,34 ± 1,12***	0,9± 0,3***£	1,52± 1,7***	0,23 0,9***#	7,63± 0,22***
	Control n"";+;.,"[b]	42,54 ± 10,08***	2,98 ± 0,23*	0,78 ± 0,16**	2,88 ± 0,64*	0,65 ± 0,15	5,84 ± 0,52***	1,92 0,13***	1,23 ± 1,3***

[a] Negative Control= dechlorinated water; [b] Positive Control= copper sulphate (0.0002 g/L); * Significantly different from negative control at P<0.05; ** P<0.01; *** P<0.001 (ANOVA); # Significantly different from Point 01 at P<0.05 (ANOVA); £ Significantly different from Point 02 at P<0.05 (ANOVA).

The current contamination of water resources as a result of anthropogenic discharges is becoming a major problem with the growth of urban areas. This change in water composition obviously has deleterious effects on the organisms that inhabit these areas, as well as on human health. Among the lethal and sub-lethal effects caused by these complex mixtures in water are fertility problems, as well as cellular, metabolic and DNA alterations (VILLELA *et al,* 2003).

A challenge associated with biomonitoring is the identification of various compounds that may be responsible for the possible adverse effects associated with exposure to environmental agents. Samples obtained from source environments are complex mixtures of organic and inorganic compounds, with thousands of individual components that can interact to produce additive, synergistic or antagonistic effects (WEAVER *et al.,* 2009).

This study generated chemical and biological data from areas exposed to many sources of pollution. Although not specific, cause and effect relationships were established, with data indicating that there may be an association between physico-chemical factors, inorganic elements and different biomonitors (BATISTA *et al.,* 2016).

Based on the results of the analysis with *A. cepa,* it can be suggested that the complex mixture of agents causing the mutagenic effect observed at points P1 and P2 of the samples collected from the Poti (ETE-Leste) and Parnaiba (ETE-Pirajà) rivers in Teresina is made up of both clastogenic and aneugenic agents, as well as causing toxicity and mutagenicity. The most significant values compared to CP were observed at ETE-Leste (Poti River) during the dry period, suggesting the presence of xenobiotics at the study sites. A more critical situation can also be highlighted during the dry period, which is to be expected due to the concentration of pollutants during this period due to the lack of rain.

6 CONCLUSION

This research highlighted the importance of carrying out environmental monitoring of water bodies that receive a high demand of treated and untreated effluents, as the chemical compounds present in these effluents can pose a danger to the entire ecosystem. Tests using plants as test organisms, including the species *A. cepa,* have proved to be very efficient for this type of monitoring and are considered excellent tools for assessing the quality of these waters. Thus, in this study, the *Allium cepa* test proved to be an efficient test organism for assessing the genotoxic and mutagenic potential of the waters and effluents of the Poti and Parnaiba rivers.

According to the results obtained in this study, we concluded that the water and effluents collected showed high concentrations of electrical conductivity, detergents and phosphorus, especially at the P1 and P3 points of the ETE-Leste, as detected by the physico-chemical analyses, in addition to the high levels of thermotolerant coliforms found in the microbiological analyses at the two sewage treatment plants, especially during the dry period, considered the most critical, indicating that the watercourses of the tributaries under study are being affected by different xenobiotics, characterizing a negative impact on the environment.

On the other hand, with the results of the *Allium cepa* test, we can infer that the treatment of the effluent from the East and Pirajâ sewage treatment plants had a genotoxic and mutagenic potential, proven by the alterations induced in the cells after exposure to the water and effluent samples, This may have been due to the release of xenobiotics from other anthropogenic sources, such as pit-cleaning vehicles, which discharge the effluent collected at this station, unlike the Pirajá sewage treatment plant, which only receives domestic sewage. The waste from these pit-cleaning vehicles possibly contains a complex mixture of mutagenic agents, which is compromising the efficiency of the treatment at the ETE-East.

Therefore, as a suggestion, it is crucial that sewage treatment plants are prepared to receive only waste whose origin and composition are appropriate to the treatment technology used, as the treatment complex in the eastern zone (ETE) does not have this characteristic, which is why it showed more inducing values of toxicity, genotoxicity and mutagenicity.

7 BIBLIOGRAPHICAL REFERENCES

ANA. National Water Agency. **Conjuncture of Water Resources in Brazil: Report 2015.** Brasilia, 2015.

ANA. National Water Agency. **Water quality indicators.** Available at: <http://pnqa.ana.gov.br/IndicadoresQA/IndiceQA.aspx>. Accessed on: Jan. 2010.

AGUIRRE-GONZALES, M.; TABORDA-OCAMPO, G.; DUSSAN-LUBERT, C.; NERIN, C.; ROSERO-MOREANO, M. Optimization of the HS-SPME technique byproducts by GC in drinking water. Jour Braz, **Chem** Soc.; 12: 2330-36. 2011.

ALMEIDA A.F.S *et al.* **UEL** v. 24. 2003. p. 87.

AL-SABTI, K.; METCALFE, C. Fish micronuclei for assessing genotoxicity in water. **Mutation Research**, Amsterdam, v. 343, p.121-135, 1995.

ALVES, R. **Thermotolerant Coliforms and Escherichia coli.** Available at: <http://www.webartigos.com/articles/28006/1/Coliformes-Termotolerantes-e- Escherichia-Coli/pagina1.html#ixzz1Pw6J1Nr4>. Accessed on: June 2011.

ALVIM, L. B.; KUMMROW, F.; BEIJO, L. A.; LIMA, C. A. A.; BARBOSA, S. Cytogenotoxicity evaluation of textile effluents using *Allium cepa* L. **Revista Ambiente & Agua**, v. 6, n. 2, p. 255-265, 2011.

ANDRADE, V.M.; FREITAS, T.R.O.; SILVA, J. Comet assay using mullet (*Mugil* sp.) and sea catfish (*Netuma* sp.) erythrocytes for the detection og genotoxic pollutants in aquatic environment. **Mutation Research,** Amsterdam, v. 560, p.57-67, 2004.

APHA (American Public Health Association, Washington). **Standard Methods for the Examination of Water and Wastewater**, 20th ed. USA: Washington, 2005.

ARAMBASIC, M. B.; BJELIC, S.; SUBAKOV, G. Acute toxicity of heavy metals, phenol and sodium on *Allium cepa* L., *Lepidium sativum* L. and *Daphnia magna* St.: comparative investigation and the practical applications. **Water Research**, v. 29, p. 497-503, 1995.

ARAÙJO, J. L. L. **Atlas escolar do Piaui**. Joao Pessoa: Grafset, 2006. p.52-57.

BARBÉRIO, A.; VOLTOLINI, J.C.; MELLO, M.L.S. Standardization of bulb and root sample sizes for the *Allium cepa* test. **Ecotoxicology,** London, v. 20, n. 4, p. 927-935, 2011.

BARBÉRIO, A.; BARROS, L.; VOLTOLINI, J.C.; MELLO, M.L.S. Evaluation of the cytotoxic and genotoxic potential of water from the River Paraiba do Sul, in Brazil, with the Allium cepa L. test. **Brazilian Journal of Biology**, Sao Carlos, v. 69, n. 3, p. 837-842, 2009.

BARBOSA, C. F. **Hydrogeochemistry and nitrate contamination in groundwater in the Piranema neighborhood, Seropédica - RJ.** 2005 (Master's thesis) - State University of Campinas, Campinas. Available at: <http:// www.bibliotecadigital.unicamp.br>. Accessed on: September 21, 2014.

BATISTA, N.J.C., CAVALCANTE, A.A.C.M., OLIVEIRA, M.G., MEDEIROS, E.C.N., MACHADO, J.L., EVANGELISTA, S.R., DIAS, J.F., DOS SANTOS, C.E.I., DUARTE, A., DA SILVA, F.R., DA SILVA, J. Genotoxic and mutagenic evaluation of water samples from a river under the influence of different anthropogenic activities. **Chemosphere,** 164, 134-141. 2016.

BELANGER, K. D, *et al.* The karyopherin Msn5/Kap142 requires Nup82 for nuclear export and performs a function distinct from translocation in RPA protein import. J Biol. **Chem** 279, 42: 43530-9. 2004.

BERNARDO, L. DI; PAZ, L. P. S. **Selection of water treatment technologies.** Sao Carlos: LDiBe, 2010. p. 868.

BIANCHI, J.; ESPINDOLA, E. L. G.; MARIN-MORALES, M. A. Genotoxicity and mutagenicity of water samples from the Monjolinho River (Brazil) after receiving untreated effluents. **Ecotoxicol Environ Saf,** v. 74, n. 4, p. 826-33, 2011.

BRAZIL. **National Basic Sanitation Plan.** Ministry of Cities. National Secretariat for Basic Sanitation. Brasilia, 2014. Available at:

<http://www.cidades.gov.br/images/stories/ArquivosSNSA/PlanSaB/plansab_texto_edit ado_para_download.pdf>. Accessed on September 12, 2016.

BRAZIL. **National Sanitation Information System (SNIS): Diagnosis of Water and Sewerage Services.** Ministry of Cities. National Secretariat for Environmental Sanitation - SNSA. 2014. Brasilia. Available at: <http://www.epsjv.fiocruz.br/upload/Diagnostico_AE2014.pdf>. Accessed on: September 10, 2016.

BRAZIL. National Environmental Council (CONAMA). **Resolution No. 430, of May 13, 2011.** Provides for the conditions and standards for discharging effluents, complements and amends Resolution No. 357, of March 17, 2005, of the National Environment Council.

BRAZIL. National Environmental Council (CONAMA). **Resolution No. 357, of March 17, 2005.** Provides for the classification of bodies of water and environmental guidelines for their classification, as well as establishing conditions and standards for the discharge of effluents, and other provisions.

BRAZIL. National Water Agency; PIAUÎ. Secretariat for the Environment and Water Resources of the State of Piaui. **Atlas of water supply in the state of Piaui - supply to municipalities with less**

than 5,000 inhabitants. ANA/SAS, Brasilia: 2004.

BRAZIL. **Law No. 6.938, of August 31, 1981.** Provides for the National Environmental Policy, its purposes and mechanisms for formulation and application, and makes other provisions. Available at: http://www.oas.org/dsd/fida/laws/legislation/brazil/brazil_6938.pdf. Accessed on: September 14, 2016.

CAMPOS, T. DE S.; ROHLFS, D.B. **Evaluation of nitrate values in groundwater and their correlation with anthropic activities in the municipality of Aguas Lindas de Goiàs.** Goiânia. PUC, 2011. Available at: <http:// www.cpgls.ucg.br/arquivosUpload/1/File/.../SAUDE/86.pdf.>. Accessed on: 02 Apr 2014.

CARITÀ, R.; MARIN-MORALES, M. A. Induction of chromosome aberrations in the Allium cepa test system caused by the exposure of seeds to industrial effluents contaminated with azo dyes. **Chemosphere,** v. 72, p.722-725, 2008.

CELINO, J. J.; CORSEUIL, H. X.; FERNANDES, M.; GARCIA, K. S. Distribution and sources of polycyclic aromatic hydrocarbons in the aquatic environment: a multivariate analysis. **REM: R Esc Minas,** v. 63, n. 2, p. 211-218, 2010.

CETESB (SÂO PAULO ENVIRONMENTAL TECHNOLOGY AND SANITATION COMPANY). **Appendix D - Environmental and Health Significance of Quality Variables.** Sao Paulo, 2013. Available at:

<http://aguasinteriores.cetesb.sp.gov.br/wpcontent/uploads/sites/32/2013/11/Ap%C3% AAndice-D-Significado-Ambiental-e-Sanit%C3%A1rio-das-Vari%C3%A1veis-de- Qualidade.pdf.>. Accessed on: June 15, 2016.

CETESB (SÂO PAULO ENVIRONMENTAL TECHNOLOGY AND SANITATION COMPANY). **Water quality variables.** Sao Paulo, 2010. Available at: <http://www.cetesb.sp.gov.br/agua/Àguas-Superficiais/34-Variàveis-de-Qualidade-das- àguas> Accessed on: June 10, 2016.

CETESB (SÂO PAULO ENVIRONMENTAL TECHNOLOGY AND SANITATION COMPANY). **Report on the quality of inland waters in the state of São Paulo 2009.** Sao Paulo: CETESB, 2009. 288p.

CHANDRA, S.; CHAUHAN, L. K.; MURTHY, R. C.; SAXENA, P. N.; PANDE, P. N.; GUPTA, S. K. Comparative biomonitoring of leachates from hazardous solid waste of two industries using Allium test. **Science of the Total Environment,** v. 347, p. 4652, 2005.

CHEN, Y.; NIU, Z.; ZHANG, H. Eutrophication assessment and management methodology of multiple pollution sources of a landscape lake in North China. **Environ Sci Pollut Res,** v. 20, p.

3877-3889, 2013.

CLAXTON, L.D., HOUCK, V.S., HUGLES, T.J. Genotoxicity of industrial wasted and effluents. **Mutation Research,** v. 410, p. 237-243, 1998.

CODEVASF. Companhia de Desenvolvimento dos Vales do Sao Francisco e do Parnaiba. **Action Plan for the Integrated Development of the Parnaiba Basin (PLANAP):** Atlas of the Parnaiba Basin. Brasilia, DF: TDA Desenho & Arte Ltda., 2006.

COTELLE, S.; FERARD, J.F. Comet assay in genetic ecotoxicology: a review. **Environmental and Molecular Mutagenesis,** v. 34, p.246-255, 1999.

COTELLE, S.; MASFARAUD, J.F.; FÉRARD, J.F. Assessment of the genotoxicity of contaminated soil with the *Allium/Vicia-micronucleus* and the *Tradescantia* micronucleus assays. **Mutation Research** - Fundamental and Molecular Mechanisms of Mutagenesis, Amsterdam, v.426, p.167-171, 1999.

DA SILVA, J. *et al.* Evaluation of the genotoxic effect of rutin and quercetin by comet assay and micronucleus test. **Food and Chemical Toxicology,** n. 40, p. 941-947, 2003.

DE SILVA, F. C; BARROS, M.A.B; VIANA, R.R; ROMÂO, N.F; OLIVEIRA, M.S; MENEGUETTI, D.U.O. Evaluation of mutagenesis caused by iron sulfate through the micronucleus test in mouse bone marrow cells. **Revista cientifica da Faculdade de Educaçâo e Meio Ambiente,** 2 (1) :13-22, 2011.

DUARTE, F. C; LAHOZ, R.A.L. Basic sanitation and the right to health: considerations based on the principle of universalization of public services. **Journal of Constitutional Studies, Hermeneutics and Theory of Law (RECHTD).** Rio Grande do Sul, v. 7, n. 1, p. 62-69, Apr. 2015.

EGITO, L.C.M.; MEDEIROS, M.G.; MEDEIROS, S.R.B.; AGNEZ-LIMA, L.F. Cytotoxic and genotoxic potential os surface water from the Pitimbu river, northeastern/RN Brazil. **Genetics and Molecular Biology,** Ribeirao Preto, v.30, n.2, p.435-441, 2007.

ENGESOFT. Engenharia e Consultoria LTDA. **Elaboration of the Conception Study, Basic Project and Executive Project of the Sanitary Sewerage System of the City of Teresina-PI,** v. 2, Volume I - Diagnosis of the Existing Sanitary Sewerage System, 2013. Volume IB - Diagnosis of the Existing Water Supply System, 2013.

EUGRIS. Portal for soil and water management in Europe. **Contaminated land.** United Kingdom, 2013. Available at: <http://www.eugris.info/eugris-soil-water encyclopedia.asp>. Accessed on: June 4, 2013.

EVSEEVA, T.I.; GERAS'KIN, S.A.; SHUKTOMOVA, II. Genotoxicity and toxicity assay of water sampled from a radium production industry storage cell territory by means of *Alliumtest*. **Journal of Environmetal Radioactivity**, Oxford, v.68, p.235248, 2003.

FATIMA, R. A.; AHMAD, M.; Genotoxicity of industrial wastewaters obtained from two different pollution sources in northern India: A comparison of three bioassays. **Mutation Research**, v.609, p.81-91, 2006.

FENECH, M. The in vitro micronucleus technique. **Mutation Research**, v.455, p.8195, 2000.

FENECH M. Cytokinesis-block micronucleus assay evolves into a "cytome"assay of chromosomal instability, mitotic dysfunction and cell death. **Mutat. Res**, v. 600, p.5866, 2006

FERNANDES, T.C.C.; MAZZEO, D.E.C.; MARIN-MORALES, M.A. Mechanism of micronuclei formation in polyploidized cells of *Allium cepa* exposed to trifluralin herbicide. **Pesticide Biochemistry and Physiology**, v. 88, n.3, p.252-259, 2007.

FISKEJO, G. The *Allium* test as a standard in environmental monitoring. **Hereditas**, Lund, v.102, p.99-112, 1985.

FISKESJO, G. The *Allium* test - an alternative in environmental studies: the relative toxicity of metal ions. **Mutation Research,** Leiden, v. 197, n. 1, p. 243-260, 1988.

FISKESJO, G. The Allium test in wastewater monitoring. Environmental Toxicology and Water Quality: An **intentional journal**, v. 8, p. 291-298, 1993.

GEREMIAS, R., BARTOLOTTO, T., WILHELM-FILHO, D., PEDROSA, R. C., FAVERE, V. T. Efficacy assessment of acid mine drainage treatment with coal mining waste using *Allium cepa* L. as a bioindicator. **Ecot and Envir Saf**. 2012; 79:116-121.

GUERRA, R. C. **Studies of the sludge generated in a biological reactor for the treatment of oil production water at the Almirante Barroso Maritime Terminal in the municipality of Sao Sebastiao, SP. With a view to its final disposal.** 126p. (Doctoral Thesis) - Institute of Biosciences, Paulista State University, Rio Claro/SP, 2009.

GRANT, W. F. The present status of higher plant bioassays for detection of environmental mutagens. **Mutation Research**, Amsterdam, v. 310, n.2, p. 175-185, 1994.

GRISOLIA, C.K.; OLIVEIRA, A.B.B.; BONFIM, H.; KLAUTAU-GUIMARAES, M.N. Genotoxicity evaluation of domestic sewage in a municipal wastewater treatment plant. **Genetics and Molecular Biology**, Ribeirao Preto, v. 28, n.2, p.334-338, 2005.

GRiSOLiA, C. K.; STARLiNG, F. L. R. M. Micronuclei monitoring of fishes from Lake Paranoà, under influence of sewage treatment plant discharges. **Mutation Research**, Amsterdam, v. 491, p.39-

44, 2001.

GROVER, i. S., KAUR, S. Genotoxicity of wastewater samples from sewage and industrial e uent deected by the *Allium* root anaphase aberration and micronucleus assays. **Mutation Research,** v. 426, 183 - 188. 1999.

GUPTA, A.K., AHMAD, M. **Assessment of cytotoxic and genotoxic potential of refinery waste effluent using plant, animal and bacterial systems.** J. Hazard. Mater. 201 and 202, 92 and 99. 2012.

HALL, J.B. DNA strand breaks and chromosomal aberrations. **Radiobiology for the radiologist,** 4 ed., Lippincott Company, Philadelphia, p. 15-27, 1994.

HANSEN, K. M. S.; WILLACH, S.; MOSBACK, H.; ANDERSEN, H. R. Parcticles in swimming pool filters - Does pH determine the DBP formation? **Chem,** 87: 241-47. 2012.

HENDRYX, M.; CONLEY, P., FEDORKO, E.; LUO, J.; ARMISTEAD, M. Permitted water pollution discharges and population cancer and non-cancer mortality: toxicity weights and upstream discharge effects in US rural-urban areas. **International Journal of Health Geographics,** v. 11, n. 9, p. 1-15, 2012.

HOLLAND, N., FUCIC, A., MERLO, D.F., SRAM, R., KIRSCH-VOLDERS, M. Micronuclei inneonates and children: effects of environmental, genetic, demographic and disease variables. **Mutagenesis** 26, 51 - 56. 2011.

HOLLAND, N. et al. The micronucleus assay in human buccal cells as a tool for biomonitoring DNA damage: the HUMN project perspective on current status and knowledge gaps. **Mutation Research,** v. 659, p. 93-108, 2008.

HOSHINA, M. M., MARIN-MORALES, M. A.Micronucleus and chromosome aberrations induced in onion (Allium cepa) by a petroleum refinery effluent and by river water that receives this effluent. **Ecotoxicol. Environ,** Saf. 72, 2090b and 2095. 2009.

HOSHINA, M.M. **Evaluation of the possible contamination of the waters of Ribeirao Claro, municipality of Rio Claro, belonging to the Corumbatai River Basin, by means of mutagenicity tests on *Allium cepa*.**52 f. Monograph. Institute of Biosciences, Paulista State University, Rio Claro, SP. 2002.

IBGE. Brazilian Institute of Geography and Statistics. **Population Estimates**

Resident in Brazil and Federative Units. 2015. Available at: <ftp://ftp.ibge.gov.br/Estimativas_de_Populacao/Estimativas_2015/estimativa_dou_201 5_20150915.pdf>. Accessed on: September 30, 2016.

IBGE. Brazilian Institute of Geography and Statistics. **2010 Census: Piaui. 2011.**

Available at: <http://www.censo2010.ibge.gov.br/dados_divulgados/index.php>. Accessed on: January 18, 2011.

IDEXX. IDEXX Laboratories. Standard Methods for Water and Sewage Testing - Colilert. 2002. Available at: <https://www.idexx.com/pdf/en_us/water/6406300l.pdf>. Accessed on: March 20, 2015.

IARMARCOVAI, G., BONASSI, S., BOTTA, A., BAAN, R. A., ORSIERE, T. Genetic polymorphisms and micronucleus formation: a review of the literature. **Mutat. Res.** 658 215-233. 2007.

JUNDI, T.A.R.E.; FREITAS, T.R.O. Evolutionary Toxicology. In: ERDTMANN, B.; HENRIQUES, J.A.P.; DA SILVA, J. **Genética Toxicológica.** Porto Alegre: Alcance, , p.101-113. 2003.

KNIE, J. L. W.; LOPES, E. W. B. **Ecotoxicological tests: methods, techniques and applications.** Florianópolis: FATMA / GTZ, p. 289, 2004.

KOVALCHUK, O., KOVALCHUK, I, ARKHIPOV, A., TELYUK, P., HOHN, B. The *Allium cepa* chromosome aberration test reliably measures genotoxicity of soils of inhabited areas in the Ukraine contaminated by the Chernobyl accident. **Mutation Research**, 415, 47-57. 1998.

KONUK, M.; LIMAN, R.; CiGERCI, İ. H. Determination of genotoxic effect of boron on Allium cepa root meristematic cells. **Pakistan Journal of Botany**, 39, 73-79. 2007.

KRISTEN, U. Use of higher plants as screens for toxicity assessment. **Toxicology** *in vitro*, United Kingdom, v. 11, p. 181-191, 1997.

KULKARNI, P.; CHELLAN, S. Disinfection by product formation following chlorination of drinking water: artificial neural network models and changes in speciation with treatment. **Scie of the Tot Envir**. 2010; 408:4202-10.

LEGAY, C.; RODRIGUEZ, M.; SERODES, J.; LEVALLOIS, P. Estimation of chlorination by-products presence in drinking water in epidemiological studies on adverse reproductive outcomes: a review. In: **Scie of the Tot Envir**. 2010; 408:456-72.

LEME, D. M.; MARIN-MORALES, M. A. Chromosome aberration and micronucleus frequencies in *Allium cepa* cells exposed to petroleum polluted water - A case study. **Mutation Research,** v. 650, p. 80-86, 2008.

LEME, D. M.; MARIN-MORALES, M. A. Chromosome aberration and micronucleus frequencies in *Allium cepa* Test in environmental monitoring: a review on its application. **Mutation Research,** v. 682, p.71-81, 2009.

LEMOS, C. T.; ERDTMANN, B. Cytogenetic evaluation of aquatic genotoxicity in human cultured lymphocytes. **Mutation Research**, Leiden, v. 467, n. 1, p. 1-9, 2000.

LEVAN, A. The effect of colchicine on root mitoses in *Allium*. **Hereditas**, Lund, v. 24, n. 4 p. 471-486, 1938.

LEVAN, A. Cytological reactions induced by inorganic salt solutions. **Nature**, London, v. 156, p. 751-752, 1945.

LIBÂNIO, M. **Fundamentals of water quality and treatment.** 3 Ed. Campinas: Atomo, 2010. p. 496.

LIMAN, R.; AKYIL, D.; EREN, Y.; KONUK, M. Testing of the mutagenicity and genotoxicity of metolcarb by using both Ames/Salmonella and *Allium* test.

Chemosphere, 80, 1056-1061.2010.

LIMAN, R.; CIGERCI, İ. H.; AKYIL, D.; EREN, Y.; KONUK, M. Determination of genotoxicity of Fenaminosulf by *Allium* and Comet tests. **Pesticide Biochemistry and Physiology**, 99, 61-64. 2011.

LIMAN, R.; GOKÇE, U. G.; AKYIL, D.; EREN, Y.; KONUK, M. Evaluation of genotoxic and mutagenic effects of aqueous extract from aerial parts of *Linaria genistifolia* sub. sp. *genistifolia*. **Revista Brasileira de Farmacognosia**, 22, 541-538. 2012.

LINS, G. A. **Environmental Impacts in Sewage Treatment Plants (STPs).** 2010. Dissertation (Master's Degree in Environmental Engineering) - Federal University of Rio de Janeiro, Rio de Janeiro, 2010. Available at:

<ttp://dissertacoes.poli.ufrj.br/dissertacoes/dissertpoli491.pdf>. Accessed on: September 15, 2016.

LIU, J. L.; Li, X.Y. Biodegradation and biotransformation of the wastewater organics as precursors of disinfection byproducts in water. In: **Chem**. 2010; 81:1075-83.

MA, T. H.; XU, C.; MCCONNELL, H.; RABAGO, E. V.; ARREOLA, G. A. The improved Allium/Vicia root tip micronucleous assay for clastogenicity of environmental pollutants. **Mutation Research**, v. 334, p. 185-195, 1995.

MACHADO, R.R.B.; PEREIRA, E.C.G.; ANDRADE, L.H.C. Temporal evolution (2000-2006) of vegetation cover in the urban area of the municipality of Teresina - Piaui - Brazil. **REVSBAU.** Piracicaba, n.3, v.5, p.97-112. 2010.

MARIN-MORALES, M. A.; LEME, D. M.; MAZZEO, D. E. C. A review of the Hazardous effects of polycyclic aromatic hydrocarbons on living organisms. In: HAINES, P. A.; HENDRECKSON, D. M. Polycyclic aromatic hydrocarbons - Pollution, health effects and chemistry. **Nova Science Publishers,** Inc, p.1-49, 2009.

MAROTTA, H; SANTOS, R. O. dos; ENRICH-PRAST, A. Limnological monitoring: an instrument for the conservation of water resources in urban-environmental planning and management. **Rev. Ambiente e sociedade**. v. 11. n 1. Campinas, jan./jun, 2008.

MATEUCA R, L. N.; AKA, P. V.; DECORDIER, I.; KIRSCH-VOLDERS, M. Chromosomal changes: induction, detection methods and applicability in human biomonitoring. In: **Biochim**. 2006; 88:1515:31

MATHEUS, C. E.; TUNDISI, J. G. **Estudo físico-quimico e ecològico dos rios da bacia hidrogràfica do Ribeirao e Represa do Lobo (Broa) J.G. Tundisi (Ed.), Limnologia e Manejo de Represas**. Série Monografias em Limnologia, Universidade de Sao Paulo, Sao Carlos, p. 419-471, 1988.

MATSUMOTO, S.T.; MARIN-MORALES, M.A. Toxic and genotoxic effects of trivalent and hexavalent chromium - a review. **Revista Brasileira de Toxicologia**, v. 18, n. 1, p.77-85, 2005.

MATSUMOTO, S.T.; MARIN-MORALES, M.A. Mutagenic potential evaluation of the water of a river that receives tannery effluents using the *Allium cepa* test system. **Cytologia,** Tokyo, v.69, n.4, p.399-408, 2004.

MATSUMOTO, S.T.; MANTOVANI, M.S.; MALAGUTTI, M.I.A.; DIAS, A.L.;

FONSECA, I.C.; MARIN-MORALES, M.A. Genotoxicity and mutagenicity of water contaminated with tannery effluents, as evaluated by the micronucleus test and comet assay using the fish *Oreochromis niloticus* and chromosome aberrations in onion roottips. **Genetics Molecular Biology**, Ribeirao Preto, v.29, n.1, p.148-158, 2006.

MAZZEO, D.E.C., FERNANDES, T.C.C., MARIN-MORALES, M.A. Cellular damages in the Allium cepa test system, caused by BTEX mixture prior and after biodegradation process. **Chemosphere**, 85, 13-18. 2011.

MENDES-CÂMARA, F. M. **Avaliaçâo da Qualidade da Agua do Rio Poti na Cidade de Teresina, Piaui**. 2011. Thesis (Doctorate in Geography) - Paulista State University, Rio Claro, 2011. Available at: http://repositorio.unesp.br/bitstream/handle/11449/104342/camara_fmm_dr_rcla.pdf?sequence=1&isAllowed=y. Accessed on: June 10, 2016.

MIRLEAN, N.; CASARTELLI, M. R.; GARCIA, M. R. D. Propagation of atmospheric fluorine pollution in groundwater and soils in regions close to fertilizer industries (Rio Grande, RS). **Quim Nova**, v. 25, n. 2, p. 191-195, 2002.

MMA. Ministry of the Environment. **Parnaiba Hydrographic Region Notebook.**

Brasilia: MMA, 2006. Available at:

<http://www.mma.gov.br/estruturas/161/_publicacao/161_publicacao03032011023605. pdf>. Accessed on: September 24, 2013.

MOLNAR, J. J.; AG BABA, J. R.; DALMACIJA, B. D.; KLASNJA, M. T.; DALMACIJA, M. B.; KRAGULJ, M. M. A comparative study of the effects of ozonation and TiO2 - catalyzed ozonation on the selected chlorine disinfection byproduct precursor content and structure. **Scie of the Tot Envir**, 425:169-75. 2012.

MONARCA, S.; FERETTI, D.; COLLIVIGNARELLI, C.; GUZZELLA, L.; ZERBINI, I.; BERTANZA, G.; PEDRAZZANI, R. The influence of different disinfectants on mutagenicity and toxicity of urban wasterwater. **Wat Res,** 34 : 4261-69. 2000.

MONTEIRO, C.A.B. **Characterization of sewage disposal in Teresina: Efficiency, Restrictions and Conditioning Aspects**. Master's dissertation. Postgraduate Program in Development and Environment (PRODEMA), Federal University of Piaui, 2004.

MONTEIRO, C.A.B. **Evaluation of Fish Farming in Treated Domestic Sewage: Zootechnical, Environmental and Quality Aspects of the Fish Produced.** Doctoral thesis. Postgraduate Diploma in Civil Engineering, Federal University of Ceará, 2011.

MORAIS, R. C. de S. **Socio-environmental diagnosis of Balneàrio Curva Sao Paulo - Teresina-Pi.** Master's dissertation in Development and Environment. Federal University of Piaui, Teresina, 2012.

MOTA, S. **Urbanization and the environment**. 4. ed. Rio de Janeiro; Fortaleza. ABES, 2011.p.49-53.

MOULY, D.; JOULIN, E.; ROSIN, C.; BEAUDEAU, P.; ZEGHNOUN, A.; OLSZEWSKI-ORTAR, A.; MUNOZ, J. F.; WELTÉ, B.; JOYEUX, M.; SEUX, R.; MONTIEL, A.; RODRIGUEZ, M. J. Variations in trihalomethane levels in three French water distribution systems and the development of a predictive model. **Mutat Res.** 2010; 28:1-12.

NIELSEN, M.H.; RANK, J. Screening of toxicity and genotoxicity in wastewater by the use of the Allium test. **Hereditas** 121, 249-254. 1994.

NUNES, E.A., DE LEMOS, C.T., GAVRONSKI, L., MOREIRA, T.N., OLIVEIRA, N.C.D., DA SILVA, J. Genotoxic assessment on river water using different biological systems. **Chemosphere** 84, 47 and 53. 2011

NUVOLARI, Ariovaldo et al. **Sewage: collection, transportation, treatment and agricultural reuse.** 1. ed., Sao Paulo: Edgard Blücher, 2003. p.35.

NUVOLARI, A. **Dicionario de saneamento ambiental**. Sao Paulo: Oficina de Textos, 2013. 336 p.

OHE, T., WHITE, P.A., DEMARINI, D.M. Mutagenic characteristics of river waters flowing through large metropolitan areas in North America. **Mutat. Res.** Genet. Toxicol. Environ. Mutagen. 534, 101-112. 2003.

OLIVEIRA, N. DE L. **Estudo da Variabilidade Sazonal da Qualidade da Agua do Rio Poti em Teresina e Suas Implicaçôes na Populaçâo Local.** Master's dissertation. Postgraduate Program in Development and Environment (PRODEMA), Federal University of Piaui, 2012. p. 113.

OLIVEIRA, L. M.; VOLTOLINE, J. C.; BARBÉRIO, A. Mutagenic potential of pollutants in the water of the Paraiba do Sul River in Tremembé, SP, Brazil, using the *Allium cepa* test. **Revista Ambiente e Agua.** v. 6, n. 1, 2011.

WHO. World Health Organization. Mental health through the prism of public health. **World Health Report 2001.** Geneva: PAHO/WHO, p.1-16, 2001.

OSTRENSKY, A. **Studies for the technological urbanization of marine shrimp farming on the coast of Paranà**, Brazil. Curitiba. 122 p. 1977 (Doctoral thesis) - UFPR, 1977.

OZKARA, A.; AKYIL, D.; ERDOGMUŞ, S. F.; KONUK, M. Evaluation of germination, root growth and cytological effects of wastewater of sugar factory (Afyonkarahisar) using Hordeum vulgare bioassays. **Environmental Monitoring and Assessment**, 183, 517-524. 2011.

PADUA, V. L. DE; FERREIRA, A. C. DA S. **Quality of water for human consumption. In: Water supply for human consumption.** Belo Horizonte: UFMG, 2006. p. 153-221.

PANTALEÂO, S.M.; ALCÂNTARA, A.V.; ALVES, J.P.H.; SPANÒ, M.A. The piscine micronucleus test to asses the impact f pollution on the Japaratuba River in Brazil. **Environmental and Molecular Mutagenesis,** New York, v.47, p.219-224, 2006.

PENTEADO, C. J. P. **Quimica Nova**, v.29, 2006.

PESSOA,C.A; JORDÂO, E.P. - **Tratamento de Esgotos Domésticos**, 4a. Ed. Rio de Janeiro, ABES, 2009.

PIVALI, R. P.; KATO, M. T. **Qualidade das Aguas e Poluiçâo: Aspectos Fisico-Quimicos**. Sao Paulo: ABES, 2005. p. 121-252.

PMT. Teresina City Hall. **Municipal Basic Sanitation Plan.** Product 02 - Diagnosis of the basic sanitation situation. Teresina, 2015.

Available at: http://semplan.teresina.pi.gov.br/wp-content/uploads/2015/05/PMSB-DIAGN%C3%93STICO-_-ATUALIZADO_ABRIL.pdf. Accessed on: April 20, 2016.

QUINZANI-JORDÂO, B. **Cell cycle in meristems. The formation of exchanges between sister**

chromatids. 1987. 276f. Thesis (Doctorate) - Complutense University, Madrid, 1987.

RANK, J.; NIELSEN, M.H. Genotoxicity testing of wastewater sludge using the *Allium cepa* anaphase-telophase chromosome aberration assay. **Mutation Research,** v. 418, p. 113-119, Amsterdam, 1998.

RIBEIRO, L. R.; SALVADORI, D. M. F.; MARQUES, E. K. **Environmental mutagenesis.** Canoas: ULBRA, 2003. p.90.

RICHARDSON, S.D. Disinfection by products and other emerging contaminants in drinking water. Tren in Anal In: **Chem**, 22:666-84. 2003.

ROCHA, M. E. S. da. et al. Preliminary Evaluation of the Pirajà Wastewater Treatment Plant - Teresina - PI.In: 21st BRAZILIAN CONGRESS OF SANITARY AND ENVIRONMENTAL ENGINEERING, 2001, Joao Pessoa. **Proceedings**... Rio de Janeiro: ABES, 2001, p. 1-6. Available at: <http://www.bvsde.paho.org/bvsaidis/aresidua/brasil/ii-110.pdf>. Accessed on: September 10, 2013.

RoSA, A.H.; FRACETo, L.F.; CARLoS, M.V. **Meio Ambiente e sustentabilidade**. Porto Alegre: Bookman, 2012. p.48-55.

RoSA, E.; AFoNSo L. **Ciência Viva**, 2008. p.77.

RUSSEL, P.J. Chromosomal mutation. in: CUMMiNGS, B. (org.), **Genetics**. San Francisco: Pearson Education inc, 2002, p. 595-621.

SANTi, G. M.; FURTADo, C.; MENEZES, R. S.; KEPPELER, E. C. Variabilidad espacial de parametros e indicadores de calidad del agua en subcuenca hidrografica del igarape Sao Francisco, Rio Branco, Acre, Brasil. **Applied Ecology**. v. 11, n. 1, p. 2331, 2012.

SEMAR. **Poti River Basin.** Piaui State water supply atlas, 2004. CD-ROM 1.

SCHMID, W. The micronuclei test. **Mutation Research**, v. 31, p. 1-15, 1975.

SHAH, A. D.; MITCH, W. A. Halonitroalkanes, halonitiles, haloamides, and N- nitrosamines: a critical review of nitrogenous disinfection byproduct formation pathways. **Envir Sci Tech**, 46 : 119-131. 2012

SILVA, I. M.; TAUK-TORNISIELO, S. M.; SANTOS, A. A. O.; MALAGUTTI, E. N. Evaluation of the water quality of the fishing grounds located in the Corumbatai River Basin, SP (Brazil). **Holos Enviromenment,** v. 12, n. 2, p. 179. 2012.

SILVA, P. P. Da.; ALVES, D. M.; NUNCIO, F.; WILLIAN, G.; NUNES, R. C.;

TORQUIM, V.; FERREIRA, D. C. **Evaluation of the Biodegradability of Commercial**

Detergents. VII UNIUBE Technology Meeting. Minas Gerais, 2011.

SMAKA-KINCL, V. *et al.* The evaluation of waste and ground water quality using the Allium test procedure. **Mutation Research**, v. 368, p. 171-179, 1996.

SOUZA, T.S.; FONTANETTI, C.S. Micronucleus test and observation of nuclear alterations in erythrocytes of Nile tilapia exposed to waters affected by refinery effluent.

Mutation Research - Genetic Toxicology and Environmental Mutagenesis, Amsterdan, v.605, p.87-93, 2006.

SOUZA, T.S.; HENCKLEIN, F.A.; ANGELIS, D.F. GONÇALVES, R.A.; FONTANETTI, C.S. Evaluation of bioremediation and biodegradation of hydrocarbons in contaminated soil, using the chromosomal aberration test in *Allium cepa*. In: **X Brazilian Congress of Ecotoxicology,** Bento Gonçalves-RS, 2008.

SOUZA, T. S.; HENCKLEIN, F. A.; ANGELIS, D. F.; GONÇALVES, R. A.; FONTANETTI, C. S. The Allium cepa bioassay to evaluate landfarming soil, before and after the addition of rice hulls to accelerate organic pollutants biodegradation.

Ecotoxicology and Environmental Safety, 72, 1363-1368. 2009.

SOUZA, T. S.; HEINCKLEIN, F. A.; DE ANGELIS, D. F.; FONTANETTI, C. S. Clastogenicity of landfarming soil treated with sugar cane vinasse. **Environmental Monitoring and Assessment,** 185, 1627-1636. 2013.

SPERLING, M. V. **Introduction to water quality and sewage treatment.** 3ª ed. Belo Horizonte. Department of Sanitary and Environmental Engineering. Federal University of Minas Gerais, 2005. 243p.

TABREZ, S., AHMAD, M. Oxidative stress-mediated genotoxicity of wastewaters collected from two different stations in northern India. **Mutat. Res**, 726, 15-20. 2011.

THEBALDI, M. S.; SANDRI, D.; FELISBERTO, A. B.; ROCHA, M. S.; NETO, S. A. Water quality of a stream under the influence of treated bovine slaughter effluent.

Revista Brasileira de Engenharia Agricola e Ambiental. v. 15, n.3, p. 302-309. 2011.

TABET, M., ABDA, A., BENOUARETH, D.E., LIMAN, R., KONUK, M., KHALLEF, M., TAHER, A. Mutagenic and genotoxic effects of Guelma's urban wastewater, Algeria. **Environ. Monit.** Assess. 187, 13 and 26. 2015.

THOMAS, P., HOLLAND, N., BOLOGNESI, C., KIRSCH-VOLDERS, M., BONASSI, S., ZEIGER, E., KNASMUELLER, S., FENECH, M., Buccal micronucleus cytomeassay. **Nat.** Protoc.4,

825 and 837. 2009.

TIPIRDAMAZ, R., G€OMÜRGEN, A.N., OLANKAYA, D., DOGAN, M.
Determination of toxicity of pulp-MIII effluents by using Allium test. **Tarim Bil. Der,** 9, 93-97.
2003.

TRÉPOS, R.; MASSON, V.; CORDIER, M. O.; GASCUEL-ODOUX, C.; SALMON- MONVIOLA,
J. Mining simulation data by rule induction to determine critical source areas of stream water
pollution by herbicides. **Comput Electron Agric,** v.86, p. 75-88, 2012.

TUCCI. C. E. M. Urban water: interfaces in management. In: PHILIPPI Jr.
Sanitation, health and the environment: tools for sustainable development. Barueri- SP: Manole,
p. 375-411, 2005.

TUCCI. C. E. M. Agua no meio urbano. In: REBOUÇAS, A. da C., et al (org.). **Aguas doces no
Brasil.** 3 ed. Sao Paulo: Escrituras, 2006. p.399-432.

TURKOGLUS. et al. Evaluation of genotoxic effects of sodium propionate, calcium propionate and
potassium propionate on the root meristem cells of Allium cepa. **Food and Chemical Toxicology**
p.1-7, 2008.

UMBUZEIRO, G. A (Org.). **Potability guide for chemical substances.** Sao Paulo: Limiar, 2012.
148 p.

UMBUZEIRO, G.D., ROUBICEK, D.A., SANCHEZ, P.S., SATO, M.I.Z. The Salmonella
mutagenicity assay in a surface water quality monitoring program based on a 20-year survey. **Mutat.
Res.** Genet. Toxicol. Environ. Mutagen. 49, 119-126. 2001.

USEPA. **Alternative disinfectants and oxidants guidance.** 815R99014. 1999.

USEPA. **Federal register,** vol 63, n. 241, 16/12/1998, Rules and Regulations. 1998.

USEPA. **National organics monitoring survey.** USEPA, Cincinnati, USA. 1978.

VARGAS, V.M.F., MIGLIAVACCA, S.B., MELO, A.C., HORN, R.C., GUIDOBONO, R.R.,
FERREIRA, I.C.F.S., PESTANA, M.H.D. Genotoxicity assessment in aquatic environments under
the influence of heavy metals and organic contaminants. **Mutat. Res.** 490, 141e158. 2001.

VENKATRAMREDDY, V.; VUTUKURU, S.S.; TCHOUNWOU, P.B. Ecotoxicology of
Hexavalent Chromium in Freshwater Fish: A Critical Review. **Rev Environ Health,** v. 24, n. 2, p.
129-145, 2009.

VENTURA-CAMARGO, B. C.; MALTEMPI, P. P. P; MARIN-MORALES, M. A. The use of the
cytogenetic to identify mechanisms of action of an azo dye in Allium cepa meristematic cells.

Journal of Environmental and Analytical Toxicology, v. 1, n. 3, p. 1-12, 2011.

VESNA, S.; STEGNAR, P.; LOVKA, M.; TOMAN, M. J. The evaluation of waste, surface and ground water quality using the *Allium* test procedure. **Mutation Research**, Amsterdam, v. 368, n. 3-4, p.171-179, 1996.

VILLANUEVA ,C. M.; CASTANO-VINYALS ,G.; MORENO, V.; CARRASCO- TURIGAS ,G.; ARAGONÉS, N.; BOLDO, E.; ARDANAZ, E.; TOLEDO, E.;

ALTZIBAR, J. M.; ZALDUA, I.; AZPIROZ, L.; GONI, F.; TARDÓN, A.; MOLINA, A. J.; MARTiN, V.; LÓPEZ-ROJO, C.; JIMÉNEZ-MOLEÓN, C. R.GÓMEZ- ACEBO, I.; PEIRÓ, R.; RIPOLL, M.; GARCIA-LAVEDAN, E.; NILUWENHUJSEN, M. J.; RANTAKOKKO, P.; GOSLAN, E. H.; POLLAN, M.; KOGEVINAS, M.

Concentrations and correlations of disinfection by-products in municipal drinking water from an exposure assessment prespective. **Eviron Res**, 114:1-11. 2012.

VILLELA, I.V.; LAU, A.; SILVEIRA, J.; PRa, D.; RoLLA, H.C.; SILVEIRA, J.D. **Bioassays for monitoring environmental genotoxicity**. In: Da Silva, J. et al (Ed.). Genética Toxicológica. Porto Alegre: Alcance, 2003. p. 147-163.

WEAVER, C.P. et al. A preliminary synthesis of modeled climate change impacts on U.S. regional ozone concentrations. **Bull. Amer. Meteorol.** Soc., 90, 1843-1863, 2009.

WHITE, P.A., RASMUSSEN, J.B. The genotoxic hazards of domestic wastes in surface waters. **Mutation Research,** Amsterdam, v. 410, p. 223-236, 1998.

WHo. **Boron in drinking water. Background document for development of WHO Guidelines for Drinking-water Quality.** 2003. Available at :

<http://www.who.int/water_sanitation_health/dwq/boron. pdf.>. Accessed on: March 31, 2014.

W.H.o. World Health organization.**International Program on Chemical Safety (IPCS)**. Environmental Health Criteria 155. Biomarkers and Risk Assessment: Concepts and Principles. Geneva. 1993.

YANG, X.; GUo, W.; SHEN, Q. Formation of disinfection by products from chlor(am) Imation of algal organic matter. **Jour of Haza Mat**, 197:378-88. 2011.

YILDIZ, M.; CIGERCI, İ. H.; KONUK, M.; FIDAN, A. F.; TERZI, H. Determination of genotoxic effects of copper sulphate and cobalt chloride in *Allium cepa* root cells by chromosome aberration and comet assays. **Chemosphere**, 75, 934-938. 2009.

Printed by Books on Demand GmbH, Norderstedt / Germany